INDUSTRY IN THE LANDSCAPE, 1700–1900

Field to factory, what next?
Is the age of British industry past?

Two hundred years of industry have transformed the British landscape. This book enables the reader to reconstruct the landscape of past industry. The authors are industrial archaeologists of national standing whose concern is to use surviving material evidence and contemporary sources to study the former working conditions of men and women. Comprehensive in coverage, the book examines fuels, metals, clothing, food, building and transport. It makes clear the tangible elements which form the basis for the recreation of past landscapes and demonstrates both their function and the context in which they should be considered.

Marilyn Palmer is Senior Lecturer in History and Archaeology at the University of Leicester. **Peter Neaverson** is Honorary Research Fellow in History at the University of Leicester. They are co-editors of *Industrial Archaeology Review* and co-authors of *Industrial Landscapes of the East Midlands* (1992).

History of the British Landscape

General editor: MICHAEL REED

Already published

The Landscape of Britain: from the beginnings to 1914
Michael Reed

Landscapes: the artists' vision
Peter Howard

Changing Scottish Landscape 1500–1800
Ian and Kathleen Whyte

INDUSTRY IN THE LANDSCAPE, 1700–1900

Marilyn Palmer and Peter Neaverson

Routledge
Taylor & Francis Group
LONDON AND NEW YORK

First published 1994
by Routledge
2 Park Square, Milton Park, Abingdon, Oxon, OX14 4RN

Simultaneously published in the USA and Canada
by Routledge

605 Third Avenue, New York, NY 10017

Routledge is an imprint of the Taylor & Francis Group, an informa business

First issued in paperback 2011

Typeset in Garamond by
Florencetype Limited, Kewstoke, Avon

British Library Cataloguing in Publication Data
A catalogue record for this book is available from the British Library.

Library of Congress Cataloging in Publication Data
Palmer, Marilyn.
Industry in the landscape / Marilyn Palmer and Peter Neaverson.
p. cm.
Includes bibliographical references and index.
1. Industrial archaeology – Great Britain. 2. Work environment – Great Britain – History. I. Neaverson, Peter. II. Title.
T26.G7P34 1994
609.41 – dc20 93-49805

ISBN13: 978-0-415-11206-2 (hbk)
ISBN13: 978-0-415-51345-6 (pbk)

Contents

Illustrations

Figures

Illustrations

Plates

Preface

Industry, by which is meant man's efforts to turn primary products into manufactured articles, is as old as his terrestrial existence, but only in the eighteenth century did the word come to mean the systematic organisation of labour for this purpose. The period from 1700 to 1900 witnessed the transformation of the British landscape on a scale never before experienced, creating scenes of horrifying grandeur and indescribable squalor which equally enthralled and disgusted those who recorded their impressions in verbal or graphic form. We can deplore the visual outcome of past industrial activity, but we cannot ignore it. Industry affected the development of many now seemingly rural areas, and often explains anomalies in the landscape: the traveller through east Leicestershire's hunting country little suspects that the elevation of the road above the surrounding fields was caused by nineteenth-century ironstone quarrying, whose extent can only be determined by studying early editions of large-scale Ordnance Survey maps.

The comparatively late date of this area of landscape history means that its students have a rich store both of visual evidence and of documentary sources to assist them in their efforts to recreate past industrial landscapes. Industrial archaeology is now a familiar term but has tended to be associated with the study and preservation of industrial monuments for their own sake. However, as in mainstream archaeology, artefacts are more and more being regarded as pointers to the past, tangible elements which form the basis for the recreation of past systems and landscapes. They prompt the industrial archaeologist to look around him for traces of sources of raw materials, water supplies, accommodation for the workforce and transport systems which together comprise an industrial landscape. Where he has the advantage over the mainstream archaeologist is that his field evidence can be verified and enhanced from a wide range of documentary and visual sources.

The purpose of this book is to describe various kinds of industrial landscape which have existed over the past two centuries, identifying the features of those landscapes which still exist in today's environment. In this way it is hoped that readers will become sufficiently familiar with these pointers to the past to be able to recognise both their function and the context with which they should be associated. The approach is archaeological and geographical rather than historical, since good accounts of the chronological development of industry in this period already exist. Industry is, however, essentially a human activity and the reason for studying the industrial landscape is to appreciate more fully the circumstances under which our forebears spent the greater part of their working lives.

Acknowledgements

The authors' appreciation of industrial landscapes has been greatly enhanced through their position as editors of *Industrial Archaeology Review* which has brought them into contact with contributors from all parts of Britain and beyond. Their membership of the Association for Industrial Archaeology and other societies has been invaluable in introducing them to different parts of Britain through attendance at meetings and conferences. It has also brought them friends who have shared their interests, especially Mary and Tony Yoward from Hampshire and David and Ann Alderton from Suffolk. Their research has been greatly assisted through the services of the Leicester University Library, the Library of the Ironbridge Gorge Museum Trust and many local archive repositories. Help with illustrations has been received from John Crompton, Mike Hodder, G.T. Knight, Pamela Moore, Miles Oglethorpe, John Powell, Paul Sillitoe and the Central Photographic Unit at Leicester University. They would also like to thank their General Editor, Professor Michael Reed, for his encouragement and John Fletcher and Janet Neaverson for their company on field trips.

1

The location of industry in the landscape

Industry and agriculture are today treated as two separate and often opposing entities. In the early eighteenth century this was not the case. These two spheres of activity necessary for human existence interacted for both economic and geographical reasons. The majority of people in Britain were still at least partially dependent on what they could grow for their survival, and industrial activity had to be carried on in association with farming and smallholding, not separated from it. In other words, industry had to go to the people, not the people to industry. Manufacturing and even mining were still usually seen as a by-employment, not as the total means of subsistence. Human muscle was still the major source of power in the early eighteenth century, and the necessary dispersal of the workforce for subsistence purposes meant that industry equally had to be dispersed. Only as agriculture began to change and produce a surplus which could support a population wholly engaged in manufacturing industry could industrial conurbations begin to develop.

Most industrial enterprises still served just a regional, if not a local, demand and their products were sold through markets, fairs and local carriers, few of which drew on so extensive a hinterland as the famous Stourbridge Fair. Daniel Defoe's *Tour Through England and Wales* of 1724 indicates how a national market for manufactured goods was already developing, but it must be remembered that he was particularly interested in what was new, not in what was commonplace. Difficulties of transport hindered the sale of goods much beyond the region, and so the production of basic necessities was replicated throughout the country. The making of boots and shoes was practised in all communities, as was brewing and malting. Agriculture did not often provide a sufficient livelihood and many families resorted to industrial by-employment. Joan Thirsk has drawn attention to the fact that the type of farming practised was important in locating a particular industry. The wood pasture regions of both Suffolk and Wiltshire, specialising in dairying, had extensive cloth industries: they were also areas populated by small freeholders or customary tenants with security of tenure, which was also true of the Yorkshire Dales. The hosiery industry of the East Midlands developed in areas of pastoral rather than arable farming, often after previously open field areas were enclosed and laid down to grass in the late eighteenth century. The wooded parishes of the West Midlands were early centres of small metalworking. Most metalliferous mining was located in the west and north of Britain where a pastoral economy predominated and there was little demand for agricultural labour, making such dual employment possible.[1]

The regional nature of the economy meant that the smallest deposits of raw materials were exploited. Coal was dug wherever it outcropped and deeper mining had already begun on all the major coalfields of Britain. Veins of metalliferous ore in remote, often mountainous, areas had been mined at intervals since prehistoric times. Pottery and bricks were made from local clays, although the quality of the earthenware produced varied considerably from one region to another. Houses were built of local materials, some of which would have been rejected in areas where better stone was available, like the chocolate-brown Greensand, known locally as carstone, which outcrops on the edge of the Fenland, or the clunch found in the chalk areas of Britain. Raw materials were not exploited to the point of exhaustion, as they were to be towards the end of our period, but utilised as and when needed by the local and regional community.

The presence of a particular industry in a given area, then, is the result of the complex interaction of a number of factors. These can be broadly divided into two categories, firstly, natural resources, particularly the extent and position of raw materials and, secondly, human resources providing the initiative and labour for the exploitation and processing of those raw materials. The general significance of these two categories will be explored in the rest of this chapter.

Natural resources

Raw materials for industry are both extracted from the earth and grown on its surface. Their production is widely dispersed across the British Isles: the small county of Leicestershire is not untypical in extracting coal, clay, ironstone, granite, slate and even some lead from within its boundaries, while its rural produce led Daniel Defoe to comment that 'the largest sheep and horses are found here, and hence it comes to pass, too, that they are in consequence a vast magazine of wool for the rest of the nation'.[2] Most regions of the British Isles have combined extractive and manufacturing industries and so developed mixed industrial landscapes.

The wide extent of mining and quarrying is made possible by the complex geological structure of Britain which means that a considerable variety of rocks and minerals is found within comparatively small geographical areas. The purpose of the industry lay in extracting ores often buried deeply beneath the earth's surface, with miners descending hundreds of feet by the end of the nineteenth century. The deepest was Williams Shaft, Dolcoath, in Cornwall, sunk to 3,600 feet by 1910. The underground aspect of metal-mining is of little interest to the landscape historian, but the nature of mineral veins does affect the surface development and must be briefly explained.

Iron is the most widespread in occurrence of all the metal deposits in Britain. From prehistoric times until the middle of the nineteenth century, most iron was obtained from nodules found in clays in the Weald, the Coal Measures in the Midlands or South Wales and the Carboniferous Limestones of the Lake District. Workings were small-scale, which suited the limited capacity of charcoal-fuelled smelting furnaces. The tremendous increase in demand for iron in the late eighteenth century and for steel by the mid-nineteenth century led first to the more extensive exploitation of the nodular ores, particularly in the Lake District, and then to the extraction of the more widespread but less productive stratified ores of the Jurassic ridge stretching from Oxfordshire up into Yorkshire. The former were generally mined, the latter quarried, and the landscape evidence reflects this difference.

Non-ferrous metals, like lead, copper and zinc, are found in veins contained in fissures resulting from faults or joints in the rock. Some rocks, particularly limestone, are naturally fissured and were filled by mineralising fluids resulting from movements or igneous activity beneath the earth's crust about 200 million years ago, which crystallised to form veins, pipes and flats which were not consistent in either quality or direction. Much of the lead ore in Britain has been derived from the limestone areas of Derbyshire and the Pennines. Earth movements resulting in the folding or faulting of the surface can, however, lead to mineralisation in a variety of rocks: copper, for example, has been worked in the Triassic sandstones of the Shropshire–Cheshire basin and lead in the slates of Wales and the Lake District. In Devon and Cornwall, the mineral veins occur along the edges of the granite masses from Dartmoor westwards to Land's End, a direct result of the igneous activity which formed the granite.

Veins, unlike seams of coal, are usually vertical or near vertical; those in Cornwall incline at about 20 degrees on average. Rake-veins, or lodes in the older literature, are the major veins which may run across country for a mile or more, and are, in essence, a nearly vertical wall of minerals which could be anything from 1 inch to 20 feet wide and of unknown depth, since drainage problems have prevented complete downward exploration. The quantity of actual ore in any vein was always uncertain, since a large part of the vein could, and frequently did, consist of 'gangue' minerals such as barytes, calcite, quartz and fluorspar. Prospecting for ore was, then, always a chancy business and was referred to as an 'adventure' by early mining companies. The irregular nature of mineralisation goes a long way towards explaining the always fluctuating fortunes of mining companies.

Coal occurs in seams or beds rather than the uncertain veins typical of non-ferrous metal-mining, and consequently one area can be worked for a long period of time, obliterating much of the field evidence for earlier periods of working. The quarrying of building stone, too, destroys evidence of the past, but its products can be seen in local buildings. The surface geology of Britain can change dramatically in just a few miles: granite, for example, is an igneous intrusion into the country rock and so in Cornwall the latter, known as 'killas', is found alongside granite in buildings in the same village. The hardness of granite meant that its use was local until that very property was found invaluable for civil engineering work on bridges and docks in the nineteenth century. It was also quarried in the Lake District, Scotland and in outcrops occurring in central England. Limestone has always been valued as a building material, both for its colour and for the ease with which it can be dressed and carved. It was extensively quarried in the Jurassic ridge stretching from Dorset to Yorkshire and in the Carboniferous Limestones of Derbyshire. Limestone slabs were also used for roofing alongside other local slates until Welsh slate penetrated the market once transport facilities were available: the output of the slate quarries of Caernarvonshire increased from under 20,000 tons in 1786 to over 90,000 tons by 1831.[3] Slate was also quarried in the Lake District.

The products of agriculture which provided raw materials for industry were even more widespread in occurrence than mineral resources. The Exe–Tees line, that classic divider of highland from lowland Britain, is significant in that the more fertile south and east boasted a greater variety of output. Yet even in Highland Scotland pure water and barley gave rise to an extensive distilling industry by the end of the eighteenth century and many small mills existed to grind oatmeal for both man and beast (Plate 1). Sheep were universal in both highland and lowland Britain, although the environment dictated which breeds would survive in a particular area and hence the type of wool available for local industry. Generally, the mountain sheep – for example, Cheviots, Herdwicks, Scottish Blackface – produced tough wool suitable for carpets and woollens, while the lowland breeds like Southdown, Romney Marsh and even the Leicester produced long staple wool suitable for worsteds and fine woollens. Cattle are less suited to hilly areas, and so leather processing was more an industry of lowland Britain, although there was an important centre in the Lake District. Flax became pre-eminently the product of Ulster, although it was grown in western Scotland and gave rise to an important linen industry there. Hemp for rope manufacture was grown in southern England, but the location of that industry was more conditioned by marine and agricultural needs than by local availability of raw materials.

The other raw materials of the textile industries, silk and cotton, introduce a third resource element into the historical location of industry – that is accessibility to ports, coastal shipping or at least navigable rivers. A series of concentric circles drawn outwards from the centre of England might be more important to the understanding of the development of industry than the Exe–Tees line. Yet improvements to rivers for navigation purposes were such that, by 1760, there were more than 1,300 miles of navigable river and much of Britain had access by this means to coastal shipping. Silk was imported from France and Italy into London, where Spitalfields became a centre for silk weaving. Cotton from India also entered London, although when America became a more important source the port of Liverpool became the chief point of entry and stimulated the Lancashire cotton industry. But location was not always the predominant consideration, since the original powered silk throwing factory was set up in Derby in the first two decades of the eighteenth century and cotton spinning was widespread later in the century around Derby and Nottingham. Here, market considerations were more important, with a flourishing hosiery industry, based on locally spun worsteds, which was beginning to diversify into silk and cotton. Only human factors can explain why

Plate 1 The restored Glendale Mill at Duirinish on the Isle of Skye. The thatched rubble-built nineteenth-century water-mill has an all-iron overshot wheel which dates from around 1900. There is a detached drying kiln, also thatched.

Macclesfield, not even accessible by canal until 1831, became an important centre for silk throwing and weaving.

Access to water-borne transport was also important for the development of extractive industry. The north-east coalfield, penetrated deeply by rivers like the Tyne, Wear and Tees, had developed far more extensively than land-locked Midland coalfields by the early eighteenth century and captured the London market. The Cornish copper-mining industry benefited from easy access to the coast, although good ports were few, with coal being brought from South Wales to power pumping engines and ore sent by return for smelting. On a more local scale, the Derbyshire lead-miners used water-power and soughs to drain their mines, even on limestone terrain, because of the difficulties of transporting coal to the hills.

The provision of water was equally important for manufacturing industry, whether for transport, power or processing. The same stream might power a corn, cotton or cloth fulling mill together with an iron forge and also be used in dyeing and bleaching works, like the River Leen on its course southwards from Papplewick to Nottingham. The iron industry was centred on the Forest of Dean and the Weald of Kent and Sussex, where furnaces and forges had access to both charcoal and water-power to drive bellows and hammers. The streams flowing off both sides of the Pennines were lined with carding and fulling mills, soon to be joined by spinning mills. But the ingenuity of man bent nature to his command. Artificial ponds and reservoirs were constructed in the most unlikely places to store water for industry and streams were tapped by leats several miles long to ensure that they remained filled. A very high proportion of the smaller areas of water in Britain were man-made, even before the days of great storage reservoirs for drinking-water which began to be built in the late nineteenth century. Artificial waterways, or canals, were also constructed from the mid-eighteenth century onwards and this finally freed industry from the locational constraints of natural rivers and streams, since coal could be taken to where it was needed. Human factors, then, while not displacing natural ones, became increasingly important in determining the location of industry in the period under consideration.

Human resources

Man has helped shape much of the present landscape of Britain and at the same time has imposed patterns of ownership on it. The Diggers may have argued in the seventeenth century that the land was a common treasury from which all should derive equal benefit, but their attempts to break down boundaries invoked no sympathy from Royalist and Parliamentarian gentry alike. Even in the twentieth century, land in public ownership amounts to only 13 per cent and a hard core of titled families still own nearly a third of the countryside. Contrary to popular belief, however, landlords generally promoted rather than hindered industrial activity because it was in their interests to do so. The earliest example of this is perhaps the 'milling soke', the right of a manorial lord to compel his tenants to use his corn mill. Attempts were made to extend this to fulling mills, but neither was strictly enforceable at law and the right had largely ceased by the nineteenth century. The practice did, however, ensure the continuance of many mills in profitable operation.

Unlike their continental counterparts, the British aristocracy and gentry generally had the power to exploit all the resources of their estates, both above and below ground. The attempts by the Crown to set up companies to work copper and brass in the interests of national defence had broken down by the beginning of the eighteenth century, as had their efforts to prevent private exploitation of ores containing precious metals, which was often the case, for example, with argentiferous lead. These victories, conducted through a Parliament dominated by the landed classes, enabled landowners to derive considerable incomes from industrial activity, from which they were not barred by custom and tradition like many of the European aristocracy. By the end of the eighteenth century, many landlords were involved in both agricultural improvement and the exploitation of mineral resources on their estates. The second Marquis of Rockingham was cited as a paragon of agricultural improvement by Arthur Young and at the same time developed the coal and iron resources of his Wentworth Woodhouse estate to the extent that the income from the mines exceeded the income from farm rents by the nineteenth century. The estate was inherited by the Earls Fitzwilliam, whose income from coal and iron rose from £4,000 in 1801 to £80,000 a hundred years later.[4] Other aristocratic industrial giants like the Dukes of Devonshire and Bridgewater, the Earls of Dudley and the Lowthers owned land in several counties and so spread their net far and wide. The Lowthers, for example, were responsible for much of the development of the southern part of the Cumberland coalfield in the early eighteenth century but also later exploited coal on estates in Yorkshire.

Many landlords extended the paternal attitudes exercised on their estates to the workforce labouring in their industrial enterprises. However, as the need for capital increased in the nineteenth century, landowners were forced to lease their mineral rights to companies of entrepreneurs and content themselves with the mineral royalties. An early example is Sir Carbery Pryse of Gogerddan in Cardiganshire, who in 1690 discovered argentiferous lead ores on his estate and attempted to work the deposit himself. By 1698 he had leased it to the Company of Mines Adventurers who already worked many of the Cardiganshire lead-mines. Over a century later, the mining entrepreneurs the Taylors, who had become mineral agents to the Duchy of Cornwall, took over the lease of Lord Lisburne's mines in Cardiganshire, developing the famous mine of Frongoch, and then expanded on to the Gogerddan and Nanteos estates, including the mines of Goginan and Cwmystwyth. By 1857 their mines were producing 70 per cent of the lead output from Cardiganshire and they had been responsible for the total transformation of the landscape by means of leats, reservoirs, mines and dressing plant.[5] The pioneering iron and tinplate works of the Hanbury family near Pontypool, established in the seventeenth century, were by the 1850s leased to the Ebbw Vale Company who greatly extended them. Capital was vital in the large-scale exploitation necessary by the nineteenth century, and generally landowners withdrew from their direct involvement in industrial expansion in earlier centuries.

Landowners with industrial empires were also keen promoters of new transport systems such as turnpike roads, horse-drawn waggonways, canals and later railways, although they tended to favour mineral lines rather than trunk railways. The Duke of Bridgewater's canal, built in the 1760s to take coal from his mines at Worsley to Manchester, is an obvious example: his brother-in-law, the Marquis of Stafford, helped promote the Trent and Mersey Canal from his estate at Trentham in the Potteries. Earl Fitzwilliam was active in the establishment of the South Yorkshire Railway, while down in Cornwall, Joseph Treffry of Fowey built an entirely new port at Par, promoted a short canal from it to serve the china clay district and built a railway line between 1840 and 1849 from St Blazey to Newquay, right across Cornwall: this was rebuilt in 1874 by Sir Morton Peto as the Cornwall Minerals Railway.[6] Even men acquiring landed estates for the first time did not neglect the potential of a good transport system. The Scottish merchant, John Christie, whose wealth was derived from the East Indian trade, utilised a horse-drawn waggonway down towards the Swansea Canal as a means of opening up the vast tracts of land on the barren Great Forest of Brecon in the 1820s.

Landowners could, however, use their parliamentary influence to oppose the building of railways and canals. New lines might promote a neighbour's industrial enterprise at the expense of their own: hence Lord Rawdon, later Earl of Moira, opposed the extension of the Soar Navigation from Loughborough to Leicester in the 1780s because it would enable Derbyshire coal to compete with the output of the west Leicestershire collieries in which he had an interest. His opposition was supported by Earl Ferrers and Earl Stamford, both of whom also owned mines and limeworks in the west of the county: Moira subsequently endorsed the proposals in 1793.[7] In Essex, Lord Petre of Ingatestone Hall received £120,000 compensation for the passage of the Eastern Counties Railway across his estate and used the money to purchase another property. By 1843, this railway company had paid out £600,000 for land purchase and compensation for its 51.2 miles of line from London to Colchester.[8] Many canals and railways paid dearly for the privilege of routing their lines across landed estates and progress was often delayed by negotiations.

Some landowners also opposed railways and canals on aesthetic grounds, not wishing public lines to traverse their private estates and spoil carefully planned vistas. Such opposition often helps to explain the otherwise unaccountable eccentricities in the routes taken by canals and railways. Lord Moira, although desperately needing the Ashby Canal in 1792 to open up markets for his new coalmines and ironworks, supported the objections of Penn Assheton Curzon to the line crossing his estate at Gopsall and interfering with springs which fed his ornamental lakes.[9] Even when the railways came, Leicestershire landowners continued to create local difficulties. Lord Harborough, the owner of Stapleford Park, forced the Syston to Peterborough Railway to be routed around his land, creating a tight curve which later had to be eliminated for safety purposes. The Countess of Bridgewater suggested that the London and Birmingham Railway be routed through her property alongside the Grand Junction Canal between Berkhamsted and Tring rather than breaking new ground between Uxbridge and Aylesbury (Figure 1). She argued that the land was 'already gashed by the canal' and that a railway beside it would make very little difference.[10] Robert Gordon of Kemble House obtained from the Cheltenham and Great Western Union Railway both compensation for damage to his amenities and an undertaking that the line should be tunnelled beneath his property and no public station should be situated on it.[11] The Duke of Wellington insisted that no station be built without his consent within 5 miles of his house at Stratfield Saye in Hampshire.[12]

Other landowners embellished their estates with elaborate viaducts or decorated tunnel portals if they did permit a canal or railway to cross their land. The portal at the eastern end of the Sapperton Tunnel on the Thames and Severn Canal faces into Cirencester Park, the home of Earl Bathurst, and is an elaborate classical structure compared with the simpler portal at the western end (Plate 2). Lord Anson added a crenellated tunnel portal to the collection of classical monuments already on his estate at Shugborough in Staffordshire. Specially designed or located railway stations were another privilege which many of them obtained. The Duke of Bedford was placated by a complete series of half-timbered cottage-style stations built to his approved design on the Bedford to Bletchley line. The elaborate station at Redmile incorporated a private waiting room for the Duke of Rutland, since the station served nearby Belvoir Castle, and the ducal arms were displayed on one of the gables. In such ways, then, landowners influenced the form and development of transport systems as well as the industries which they served.

Estate boundaries have played a major role in shaping the industrial landscape. Vigorous seigneurial initiative could lead to the development of one estate while an adjoining one might be neglected by a more lethargic landowner. The division of the countryside using man-made rather than geographical boundaries could also lead to similar industrial enterprises

Figure 1 A transport corridor in Hertfordshire. The Grand Junction Canal in Berkhampsted faced new competition when the London to Birmingham Railway passed through in 1838. The station has no platform but is lit by gas and was provided with a private room for the use of the Earls Brownlow (From Thomas Roscoe's illustrations of the London and Birmingham Railway, by courtesy of the Elton Collection: Ironbridge Gorge Museum Trust).

developing independently, whereas, from an economic point of view, they might have flourished better as a combined unit. The distribution of iron furnaces in South Derbyshire in the late eighteenth century bore more relation to estates occupied by the Hastings, Burdetts and Ferrers than to purely geographical considerations. Those three families and the Harpur Crewes of Calke Park each also worked limekilns based on the same two inliers of Carboniferous Limestone. In Dimminsdale, the Crewes and Ferrers operated two sets of kilns separated by only a small stream, which marked their estate boundary.

Personal factors like these have helped to create the rich variety of our industrial landscape, but it is not only the great landowners who have been responsible. People who left villages and hamlets for one reason or another often squatted on common lands and eked out a precarious existence by small-scale mining and quarrying. Such settlement can be detected from estate maps and can still be seen on the ground; it is common in Shropshire and on the edges of the Midlands coalfields, where scattered houses in isolated plots rather than nucleated settlements are still the norm.[13] Squatter communities provided a mobile source of labour which could play a major role in local industrial development. Equally, landowners themselves could encourage settlement on new ground to bring it into cultivation. The miners

Plate 2 The ornate eastern portal of the Sapperton Tunnel on the derelict Thames and Severn Canal. The 3,817 yard tunnel, 15 feet wide and 15 feet high, was begun in 1783 and took six years to complete. It was described by Robert Whitworth, the canal's engineer, as 'much longer and . . . wider than any which has yet been done'. This pedimented Classical portal is located in the Earl of Bathurst's Cirencester Park and has a space for an inscription and niches for statues, neither of which was ever used.

who rented cottages from Earl Fitzwilliam at Elsecar in South Yorkshire were also able to rent an additional 300 yards of ground for potato growing.[14] The Bassets of Tehidy, whose land embraced the Cornish mining parishes of Camborne, Illogan and Redruth, derived most of their wealth from copper and tin royalties as well as from leasing land to men seeking employment in the mines. Francis Basset, Baron de Dunstanville, whose monument adorns the heights of Carn Brea, leased 3 acres of land for three lives to these miner–smallholder tenants on condition that when the first 3 acres were cultivated, another 3 acres would be added. In this way he not only provided homes for 2,000 people between 1798 and 1842 but also brought 400 acres of wasteland into cultivation, giving these mining areas their characteristic settlement pattern of scattered smallholdings.[15]

In some areas, the rights of ordinary men to exploit natural resources were both recognised and regulated. The mining customs in the King's Field in Derbyshire allowed any man to search for lead ore without hindrance from the landowner, and claims had to be registered with the Barmaster: disputes were heard by the Barmote Court. Such rights permitted the small-scale mining of lead to continue along the rakes, resulting in a hillocky landscape of waste heaps which is more characteristic of Derbyshire lead-mining than the Taylors' highly capitalised Magpie Mine near Sheldon. In Cornwall, tinners had the right to divert streams, cut fuel and prospect for ore like the Derbyshire miners: their disputes were settled by the Stannary Courts and they paid for their privileges by a proportion of their product after smelting. The stream-workings on Bodmin Moor and Dartmoor are as characteristic of the Cornish mining landscape as the more obvious engine houses which date from the nineteenth century. The Free Miners of the Forest of Dean also retained the right to mine freely for coal and many of their small adit workings are still present in today's landscape.

It tends to be assumed that the rural industry consisted mainly of women spinning to add to the family income, but weaving, working the knitting frame and cutting and sewing leather were largely male occupations. In Kendal, for example, Arthur Young saw farmers and labourers bringing their yarn to the weekly market: he reckoned that there were from 1,000 to 1,300 wool-spinners and 500 weavers employed in the vicinity of the town, much of the yarn being used in the manufacture of the stockings for which Kendal was famous.[16] In some areas, however, women did attempt to supplement the wages their husbands earned in agricultural labour by industrial by-employment. Such were the pillow lace workers of the South Midlands: Arthur Young found 500 women and girls making lace in Bedford in 1768, earning an average of 8d. to 10d. a day.[17] The silk industry of Macclesfield employed women to make silk-covered buttons, fashionable wear on luxury garments in the eighteenth century. Women did, of course, card wool, spin thread, wind bobbins and make up stockings in the other textile industries, but as part of the family unit of production rather than as sole wage-earners in their own right.

In the course of the eighteenth century, the increasing use of water-power did lead to some centralisation of the industry but not as rapidly as is often thought. Fulling had been water-powered since the thirteenth century and carding was added, both carried out in small rural mills which served domestic workers in their own region. Powered throwing in the silk industry and spinning in the cotton and woollen industries followed, again located in rural areas and still leaving colonies of handloom weavers pursuing their traditional home-based industry. Once power, particularly steam power, began to be applied to processes, it dictated its own momentum: hours of work were determined by the incessant running of the machine, not the worker, and combined industrial and agricultural occupations ceased to be the norm. This was, however, a long drawn out process lasting two centuries. The spectacular growth of industrial towns and of industrial regions such as Coalbrookdale, much remarked on by contemporary travellers, should not blind us to how much industry continued on its rural, scattered, even part-time, basis until well on in the nineteenth century.

Analysing the landscape

Once established in a particular location, industrial activity tends to remain even after the reasons for its original establishment have long since disappeared. Northamptonshire is still the premier area for the manufacture of footwear in Britain, lace is still made in Nottingham, silk in Macclesfield and hosiery in the East Midlands. Only recently has cotton spinning declined in the Derwent Valley of Derbyshire due to the general depression in the textile industry. In Coalbrookdale, a working foundry is situated just west of Abraham Darby's furnace of 1709, maintaining an ironworking tradition going back nearly 300 years. Where the locational factor was raw materials which have subsequently been exhausted, the original extractive industry will have ceased to exist, but similar industries based on raw materials imported into the region may well have taken its place, as in the Potteries. A particular industry generates its own infrastructure of suppliers, markets, skilled labour, secondary finishing industries, transport facilities and housing for the workforce which often results in continuity. For this reason, present-day industrial landscapes need to be taken apart by landscape historians and the different layers of development determined. A stratigraphical approach of this kind gives meaning to the term industrial archaeology for, undoubtedly, no surviving industrial monument should be studied in isolation from its physical environment or its historical function. The complexity of present-day industrial areas makes the task of the industrial archaeologist a difficult one, particularly when so much of the superstructure of industrial sites is swept away with little trace. However, the fact that there is so much documentary evidence for this kind of archaeology perhaps outweighs the disadvantages of the ephemeral nature of much of the industrial landscape: maps, plans, paintings, engravings and even photographs, as well as the written word, enable these landscapes to be recreated.

The adoption of a six-point plan may well be useful in attempting to understand the historic industrial landscape. Each of the six factors considered below may have played a part in the establishment of industry in a particular location, either in its origins or subsequent development. They cannot, in practice, be treated independently of each other: a negative answer to one of the questions prompts a closer look at the others and the sequence of events may thus be determined.

Source of raw materials

When industry was first established on a particular site, were minerals or the products of agriculture responsible for its presence? If the raw materials were non-local, or imported like silk or cotton, other factors were clearly of more importance. In general, the specific location of extractive industry is governed by the presence of raw materials, but these have proved to be less vital in determining that of manufacturing industry, particularly textiles.

Processing plant

What buildings and other structures were necessary to process the raw materials? In this category come the industrial monuments which so often prompt studies of sites in the first place – the furnaces of the iron industry, kilns for bricks, pottery and glass, dressing floors for mineral working, mills and factories for textile and leather manufacture. Most early industrial structures were built in the functional tradition, that is in a style determined by the function they were to carry out: warehouses have large spaces for the storage of goods, loading doors, small and often barred windows to prevent theft; mills built before the introduction of gas or electric lighting were rarely more than 30 feet wide with numerous windows to admit light, and were multi-storey to permit the transmission of power from a central source. Careful and informed study of surviving structures can, then, determine the type of industry first carried out on a particular site even in the absence of documentary evidence. But some industrial buildings can be reused for other purposes and their subsequent functions also need careful study.

Power sources

In the eighteenth century, power sources and fuel were often the prime factor in industrial location. All over Britain, rivers and small streams were harnessed time and again to provide power for forges and mills. Steam power did not govern the location of industry in this way, since coal, unlike water-power, could be transported, but local availability of coal often determined whether steam could be used or not. Oil and gas engines could also be taken where they were needed but were not as powerful as large steam engines and by no means replaced steam as a power source in the second half of the nineteenth century. Only when electricity became widely used did availability of power cease to be a prime locational factor for industry, and even then some industries have been situated where it was most economic to make use of hydroelectric power, as in the case of the aluminium-smelting works at Kinlochleven. The generation of power often led to the creation of major landscape features such as leats, reservoirs and engine houses which have remained long after the power source has ceased to be utilised.

Secondary industry

The presence of a particular industry generated others either concurrently or subsequently. In the case of extractive industries, this could be because other minerals were discovered in shafts sunk to reach the one previously sought. Mining for coal often brought up clay and even iron ore out of the Coal Measures, and so brick and pottery industries and iron furnaces shared the pit-head sites. In manufacturing industry, the primary industry created a need for ancillary processes. In the hosiery industry, for example, bleaching and dyeing were necessary adjuncts and premises for them were established in riverside locations. Packaging was also important, and box-making and printing developed, now often surviving as separate industries in their own right. An experienced labour force might also attract different industries needing similar skills. When the East Midlands hosiery industry became largely mechanised from the 1870s and the workforce needed was female rather than male, the skills learned by the framework knitters in setting up and repairing their frames was put to good use in the new engineering industries attracted to the area, which superimposed a new layer on the industrial landscape.

Accommodation

Where industry was carried on as a by-employment to agriculture, changes had to be made to the home to accommodate loom or knitting frame. Light was of supreme importance, and additional windows were often installed. Separate workshops were also built on to existing houses. Alterations like these can be still seen in the East Midland and Pennine textile areas. In other cases, the workforce had to be attracted to the industrial sites and suitably housed. This happened in the extractive industries, where barracks for miners and quarrymen were often built in very isolated spots in the eighteenth century and entire villages on nineteenth-century coalfields. Water-powered mills and forges, too, often had to occupy previously deserted valleys and whole new communities were created. The construction of canals and railways both generated new settlements and altered the structure of existing ones through which they passed. Canal settlements often grew up where the canal passed under a turnpike road or where transshipment was necessary. Terraces of railway housing form a distinctive feature of the landscape alongside many railway lines, even the isolated ones which are now disused. If canals or railways passed close to rather than through a town or village, the pattern of housing often shifted towards them, abandoning the original centre of settlement. Finally, the transformation of industry into a full-time occupation, freed from its dependence on part-time agricultural labour or water-power, resulted in the spectacular growth of industrial towns during the nineteenth century. The pattern of accommodation for the workforce is, then, a good indicator of the stage of development of a particular industry and an essential factor in the study of the industrial landscape.

Transport

Was a particular industrial site chosen because of its proximity to means of transport or did new transport systems have to be built to move the products to the consumer? The relationship of transport networks to the location of industry is a very complex one. The scattered nature of industrial activity in the early eighteenth century resulted in an immense network of packhorse tracks and carrier routes, many of which are still detectable in the landscape of areas such as the Pennines and Wales. Contemporary travellers provide evidence about the deeply rutted roads resulting from heavy industrial traffic. Primitive railway systems along which horses drew waggons on rails were established to bring coal, stone and other bulky materials down from the mines and quarries. The first major canals provided cross-country routes from coast to coast, but the Bridgewater Canal and later canals were built specifically to transport heavy goods from their source to the main network, and waggonways were also constructed to the canal head. Railways followed a similar pattern. Once built, however, they encouraged the development of industry alongside.

In general, extractive industries generated their own transport systems because their location had to be a very specific one. Manufacturing industry, dependent on water-power, was also responsible for the direction taken by packhorse tracks and early canals. During the late eighteenth and nineteenth centuries, the establishment of a national transport system began to dictate the location of industry rather than itself be created by it. There are, of course, many exceptions to this general rule and every historic industrial landscape needs careful examination to determine the relationship between its location and the means of delivering finished goods to the consumer.

These six factors, when working together, create an industrial landscape which may be likened to a palimpsest in which one layer of development is superimposed on another. In some cases, the visual evidence for a previous activity may have been incorporated in the next development, as when a stream ceased to be a source of power once a steam engine was installed, but continued to provide water for finishing processes. In other cases, areas could be completely cleared of buildings to make way for a new mill or factory, and only the study of successive large-scale maps, if they exist, can enable the previous phase of development to be studied. Nor does the area necessarily retain any industrial activity in today's landscape: ironstone quarrying in east Leicestershire and in Lincolnshire has left only a pattern of fields lower than the roads passing over them as witness to the layer of ironstone which has been removed.

Students of historic landscapes are interested in the origins and subsequent development of the changes man has wrought on the face of the earth. In the last two centuries, industrial activity has probably been the major force for change and is therefore worthy of study, unattractive as that change may often be. This book sets out to examine clues in the landscape which may enable the reader to appreciate the impact of industry upon it and to demonstrate the diversity and richness of Britain's industrial past.

2

Providing the necessities of life

Civilisation really began when man was able to lead a settled existence by growing rather than hunting his means of subsistence. Although many changes had taken place in Britain by 1700, a large majority of people still grew a proportion of their own food. In the course of the eighteenth century, however, the growth and redistribution of population was accompanied by increased efficiency in agriculture which ensured that the town-dwellers and full-time industrial workforce could be fed. The number of people living in England and Wales rose from the 5.5 million estimated by Gregory King in 1688 to nearly 9 million in the first census of 1801 and to 32.5 million by 1901. In 1851 the number of people living in towns of over 5,000 population was for the first time equal to the rural inhabitants, but by 1901 the proportion had increased to three to one. Thus the number of non-growers increased rapidly, yet by dint of increasing agricultural productivity and imports, there was no famine.

Britain's extensive coastline had for long enabled the produce of different areas of the country to reach London, but the improvements to roads and rivers in the eighteenth century, followed by canals and railways, greatly increased the circulation of foodstuffs to both regional and national markets. But they were not all home-grown. Whenever political circumstances permitted, imports of food increased both in quantity and variety. The middle and upper classes spent a growing proportion of their dis-posable income on the purchase of imports like tea, cocoa, coffee, meat, sugar and tobacco. The evidence for these developments in the processing and distribution of the basic necessities can be detected in the changing forms and distribution of mills, factories and warehouses in the landscape.

The milling industry

For the majority of people of Britain in 1700, bread was the staple food and continued to be so until well into the nineteenth century. This is reflected in the number of cornmills, both water- and wind-powered, which were still a feature of most parishes. The Domesday survey of 1086, although not completed for the whole country, recorded a total of some 6,000 water-mills, mainly in the east and south of the country and occasionally less than a mile apart. In the medieval period, the distribution of mills was governed more by the right of the lord of the manor to insist that his tenants ground their corn at his mill than by geographical location. By 1700 this seigneurial right, known as the 'milling soke', had declined, although many mills, particularly in the north of Britain, still charged a toll in kind for the milling process. This was commuted to a cash payment by Act of Parliament in 1796. Northern mills concentrated on the cheaper grains such as rye, barley and oats, brought to them by the individual farmer. In the richer, densely populated south and east of England, wheat for white bread flour was the most popular grain and was bought by corn merchants for milling and resale on a commercial scale.

Water-mills still predominated over the more recently introduced wind-mills. In parts of Scotland, particularly the crofting communities of the islands, the horizontal water-wheel or Norse mill was still in operation. These tiny mills served the individual farms or hamlets which were the characteristic settlements in the area. The directly driven single pair of stones required a small supply of water and the same stream could be used to power a series of wheels by judicious construction of lades or leats, as at Sandvaat on the north-western coast of Lewis. Many continued in use into the twentieth century and their characteristic circular or oval rubble stone walls, often complete with millstones, remain but usually devoid of any reusable wooden parts. A restored example may be seen at Dounby in Orkney while the ruins of Shawbost Mill on Lewis, restored in the 1960s, indicate the fragile nature of these structures. Their distribution reflects the continuity of a subsistence economy in these isolated parts of Britain.

Where nucleated settlement predominated, as in lowland Scotland and much of England, the village or manorial mill served the community. The potential of a stream or river was harnessed over and over again along its course. The type of wheel employed was influenced by the river gradient, overshot wheels predominating where there were high falls of water as on the Pennine slopes and undershot on the sluggish rivers of East Anglia. The River Sence in west Leicestershire provides an illustration of a mix of wheel type in the repeated use of a small river to power a series of wheels, overshot in the upper reaches changing to breastshot on the lower gradients. While most mills were situated on a natural site, others made use of artificial features such as leats, weirs, storage ponds or tail-race tunnels to create an additional head of water or increased working capacity. Where rivers were used for navigation, a compromise had to be reached with millers whose water supply was threatened by locks and staunches. This was achieved by bypass cuts with locks and overflow weirs as is clearly shown in Constable's paintings of Flatford Mill on the River Stour in Suffolk. The diversion of rivers into water storage reservoirs and for urban flood prevention schemes during the nineteenth century deprived some millers of their supply and they were forced to convert to an alternative power source or close. The miller at Quorndon in Leicestershire was paid £2,750 in 1893 when the construction of Swithland Reservoir impounded his water supply.

If the sites of mills varied, so too did the forms of the buildings. Many early mills were small single-storeyed buildings with an external vertical wheel driving a single pair of millstones through gearing. These were often increased in size during the eighteenth century as more efficient wheel design or improved water storage capacity enabled multiple sets of stones to be driven through a second set of gearing. Additional height was also needed when the introduction of grain cleaning and flour dressing machines, together with storage hoppers, required a further floor. In addition to internal sack hoists, many mills sprouted a lucam or external covered platform complete with hoist for lifting grain to the top floor. Those serving a commercial market needed a granary alongside to provide additional storage. In northern England and Scotland, a drying kiln was a common addition to mills because of the damp climate. It was an essential element when oats were processed, as was increased floor space to accommodate additional stones for shelling the oats. The National Trust for Scotland preserve a good example of a stone-built drying kiln with pantiled roof at Preston Mill in East Linton (Plate 3). At the Percy family's Boot Mill, which served the manor of Eskdale in Cumbria, the drying kiln was integral with the mill and the three pairs of stones were driven by external tandem water-wheels.

Of all industrial buildings, water-mills are probably the best indicators of the vernacular building tradition since many of those surviving were built before canal or railway transport enabled the widespread distribution of common brick and Welsh slate. Areas with good building stone used this for their mills, which range from the rough-built random stone mills found in Scotland and the Pennines to the elegant dressed limestone mills of the Cotswolds. In areas lacking building stone, such as East Anglia, timber framing clad with weatherboarding over brick foundations was more usual and the mills here were usually large, as befitted the granary of England. Along the once navigable River Bure in Norfolk, a chain of eight mills ranged from Briston to Coltishall, each with a disused lock alongside. The Midland clay areas made considerable use of local brick in often

Plate 3 The rubble stone and pantiled water-mill at Preston, near East Linton, has been restored by the National Trust for Scotland. The external wood and iron low breastshot wheel drives two pairs of millstones. There is a separate circular grain drying kiln complete with roof vent.

unpretentious buildings, such as Claybrooke Magna in south Leicestershire and Hoveringham in Nottinghamshire. The attractive nature of water-mill buildings has meant the conversion of many to other uses, but close study will often reveal the history and layout of the original site.

Where a suitable tidal range and long coastal inlet coincided, mills making use of the rise and fall of the tide could be built. These were situated alongside a dam constructed across the inlet in which a sluice allowed the incoming tide to be ponded. The stored water was then released through the wheel on the ebb tide when the head was adequate for milling. Tide mills were vulnerable to wind and storm damage and their millers had to work unsociable hours. Nevertheless, they were still being constructed in the eighteenth and nineteenth centuries and can still be found around the south and south-west coasts of Britain. A particularly good example is in the care of the Pembrokeshire Coast National Park beside Carew Castle. A mill on this site was leased from Queen Elizabeth I for 10 sovereigns per annum in 1558 and the site has been in continuous use until 1937. Working tide mills operate on The Solent at Eling in Hampshire and on the Deben Estuary at Woodbridge in Suffolk (Plate 4). Often, however, only the remains of a breached dam betray the former presence of a tide mill, as at St Anthony Passage on the River Tamar.

Plate 4 The timber-clad Woodbridge tide mill on the Deben Estuary in Suffolk which, along with the former granary on the right, has been restored. The present mill is a rebuild of 1793 and has an 18-feet diameter undershot wheel driving four pairs of millstones.

The windmill came later to the milling scene than the water-mill, the earliest examples dating from the twelfth century. By 1700 the most common type was still the post mill, where the whole mill body was turned manually into the wind using a tailpole. As the structure needed to be light, these mills were built of wood, rarely had more than two pairs of millstones and consequently only had a limited capacity. Their distribution was widespread, but the mills themselves, by their very nature, were susceptible to damage by gale and fire and were frequently rebuilt. By the end of the eighteenth century they were being replaced by the sturdier brick- or stone-built tower mill, where only the cap was turned into the wind either manually or by means of the newly-invented fantail. The American Louis Simond, writing in 1810 of his travels between Berwick and Alnwick in Northumberland, remarked on the large numbers of windmills on farms: 'each of these mills has a small windmill, or rather wind wheel, behind, to work the cap round to the wind; not as in France, by means of a long lever, or tail, moved round by the miller to suit the wind'.[1] Multiple floors and increased power enabled more sets of millstones and dressing machinery to be introduced, while the lower floors could be used for storage. The tower mill reached the zenith of its construction in the first three decades of the nineteenth century and several hundred were built in the eastern half of the country. Several were built by subscription in protest against the monopoly of local millers and the high tolls they charged. Most subscription mills date from a period of high bread prices during the Napoleonic Wars. An Anti-Mill Society was formed in Hull in 1795, which built the five-sailed Anti-Mill in the town two years later. North Leverton Mill in Nottinghamshire, which began life as a subscription mill in 1813, is still working commercially. This mill is also unusual in that the tower has been heightened with a cylindrical section superimposed on the original conical tower. Urban mills, built near tall buildings, also needed additional height as can still be seen in surviving towers in Lincoln, Sleaford and Wisbech. The smock mill, such as the impressive example at Cranbrook in Kent, was also a means of raising height without increasing weight by superimposing a wooden tower on a brick base. An unusual example of this is the smock mill mounted on the roof of a brick- and flint-built barn at West Blatchington, near Hove in Sussex, which attracted John Constable to paint it during one of his summer visits to Brighton in the 1820s (Plate 5).

The first steam-powered mill, the Albion Mill in London, began work in 1786 with a Watt beam engine driving twenty pairs of millstones. Its output was obviously much greater than even the new tower mills and its introduction was therefore resisted by the milling fraternity. The Albion Mill was destroyed by fire in 1791, probably to the secret joy of many traditional millers, but over the next fifty years many more steam mills were erected, particularly in or near growing towns. These mills served a much larger market than the old parish mill and often processed grain brought in from outside the region. Steam mills were therefore often built alongside navigable rivers, canals and railways and usually constructed of brick, since the transport system which brought the grain brought the building materials as well. Examples of riverside mills may still be found on the Nene at Wellingborough, where Whitworth's Victoria Mill is still working, and on the Wey Navigation in Surrey where the striking Coxe's Lock Mill processed grain imported through London docks, but has now been converted into luxury apartments (Plate 34). Canalside examples are the three similarly named Albion Mills at Walsall, Worcester and Worksop, while rail-served mills survive at Chesterfield in Derbyshire and Mellis in Suffolk.

Although the introduction of the steam mill did cause the closure of some water- and windmills, particularly in and near to towns, it did not result in the overnight disappearance of the older forms of power. Coal for steam engines had to be purchased and brought to the mill, whereas water and wind were free, if not so reliable. Steam mills were built at the very time that the windmill had reached its optimum efficiency: new tower mills were being erected and many replaced smaller post mills. In Hull, for example, of the ten windmills in 1775, eight were post mills grinding both corn and oil seed, but, by 1831, there were thirty-three mills in the same

Plate 5 The wooden horizontal boarded smock mill at West Blatchington, near Hove in East Sussex, now belonging to the Hove Borough Council. The mill was erected around 1820 on a square brick and flint base raised through the existing farm buildings. The mill was painted by Constable in 1825 and worked commercially until 1897.

area, all of them tower mills. Over the same period the number of steam-powered mills rose to eleven, two grinding corn, six grinding oil-seed and the rest having other uses.[2] In some rural locations, a steam mill was built to supplement an earlier water-mill, as at Wickham Market in Suffolk where a steam roller mill was erected in 1869 on the opposite side of the River Deben to the earlier weatherboarded water-mill. Even allowing for the greater output of the steam mills, the older mills could hold their own in terms of profit. Some compensated for the unreliability of their power source by adding steam engines, later gas and oil engines or later still electric motors, as forms of auxiliary power. Except in the case of electrical drive, separate engine houses and chimneys were erected, although the only evidence is often the presence of an external drive pulley which could be driven from an endless belt by a portable engine. In the case of windmills, this was often added after the sails were damaged beyond economical repair.

Shortage of wind or water often led to a miller working both a wind- and a water-mill, one 'helping out' the other. At South Luffenham in Leicestershire, a water-mill on the River Chater was supplemented by a tower mill erected on nearby high ground in 1832: this went out of use in the 1890s when a multi-storey steam mill was built beside the main line railway, but the water-mill continued to grind until 1948, when the leat carrying water to the wheel was damaged and never repaired. A steam engine had been added, probably when the windmill went out of use, and was later replaced by an oil engine which 'helped out' the original power source, enabling the water-mill to survive despite the nearby steam mill. It was unusual for water- and windmills to be directly adjacent to each other and provide alternative sources of power for the same set of machinery because the latter were usually set on high ground away from valleys to catch as much wind as possible. In Norfolk, however, where wind was nearly as constant a factor as it is in Holland, some combinations did exist. At Little Cressingham there is a tower mill containing two groups of millstones on different floors, one group once driven by wind and the other by an adjacent water-wheel powered by the Watton Brook. Here, too, a portable steam engine provided an additional source of power and an oil engine was added when the windmill ceased to operate in 1916. At Burnham Overy a tower mill was superimposed upon a half-timbered water-mill on the River Burn, a not unusual way of adding windmilling capacity.

Two other factors were to influence mill type and location in the second half of the nineteenth century: one was the increasing imports of both grain and flour after the repeal of the Corn Laws in 1849 made both free of duty, and the other was the introduction of roller milling from eastern Europe. Hard wheat from America and eastern Europe could not be processed satisfactorily with millstones, but was more suited to the 'high milling' process of gradual reduction using rollers which produced three times more fine white flour than the single passage through stones of the British 'low milling' system. White flour was much in demand by the consumer, and in the 1880s many large mills switched to roller-mill production. This required increased power to drive a series of roller mills at a higher speed than had been needed for millstones and also to drive the necessary purifiers for middlings separation.

Some small country water-mills did convert to roller milling by replacing the water-wheel with a turbine. At Odstone in Leicestershire, a complete new roller milling system was installed in addition to the existing millstones, which necessitated the addition of a new storey to the mill. A few years later in 1902, the water-wheel was replaced by a turbine to provide sufficient power and speed to drive the new machinery. In a windmill tower, however, restrictions of space meant that a series of roller mills could rarely be accommodated, something which was a contributory factor to the rapid demise of wind-power. The main effect on the landscape of hard wheat and flour imports was at the port of entry, where grain and flour silos were erected for storage and large multi-storey roller mills, first steam- and then electrically-powered, were built. At Gloucester Docks, the Victoria and Britannia grain warehouses were built in the 1850s, after the Corn Law repeal, together with the City Flour Mills which are still operating. The Albert Mills were converted from a warehouse in 1869, finally closing in 1977 (Plate 29).[3] Beside the Floating Harbour at Bristol, the 1871 Welsh Back granary stands ten storeys high, in brash Bristol Byzantine style, with its loading facilities concealed within its arches. On Merseyside, at Birkenhead Docks, there are extensive ranges of grain silos and roller mills belonging to Spillers fronting Dock Road and dating from the late nineteenth century (Plate 6). Their massive size is testimony to the soaring imports of grain and the change in scale of corn milling since the early eighteenth century.

By contrast, baking bread remained on a largely domestic basis until the 1870s. Bread ovens were a common feature of the kitchen ranges of the large country house and generally of homes in the north of England because of the availability of wood faggots for fuel. Communal bakeries were more usual in the south of England where fuel was scarce. Some water-mills had bakeries attached to them, as at Carlton-in-Lindrick in Nottinghamshire and Rearsby in Leicestershire. The introduction of the steam tube oven in the mid-nineteenth century and mechanised kneading machines brought mass production to baking by 1900, and steam bakeries could be found in most towns. The Co-operative movement was responsible for the construction of many urban bakeries. A good example survives at Kettering in Northamptonshire, with its elaborate 'KICS' tie bar ends and associated terraced housing bearing the motto 'Unity is Strength'. The baking of biscuits for use at sea was a long-standing industry and mass production methods were used in the Royal Dockyards from the early nineteenth century. Rennie's granary, mill and bakery at the Royal William Yard at Stonehouse, Plymouth,

Plate 6 The massive scale of these roller flour mills, heavily reinforced by tie bars, can be appreciated. They were built at Birkenhead on Merseyside beside the West Float Dock especially to process imported wheat.

which still survives, was completed by 1835 and consumed 1,000 sacks of flour per week.[4] The important firm of Huntley and Palmers was founded in 1841 by two Quakers and twenty years later was producing over 100 varieties of biscuits. The firm eventually employed nearly one-quarter of the working population of Reading, but sadly most of its extensive factory alongside the Kennet and Avon Canal has now been demolished.

As well as massive grain stores, a feature of coastal and inland ports was the oil mill which crushed rape and other plant seeds to extract oils for lighting and other purposes. This crop was often grown to condition newly enclosed or drained fenland, although some seed was imported, particularly from Holland. The seed was crushed beneath edge runners in buildings similar in structure to corn mills, driven by horse-, wind- or water- power. The pulp was then bagged up and pressed to extract the oil and the solid residue formed the basis of cattle cake. The decline in demand for lighting oils following the introduction of gas was compensated by the increased use of edible oils for making margarine from the 1870s. This soft oil production was later augmented by the so-called hard oils from palm kernels and coconuts which were also used for soap manufacture and chemical production. Wisbech, in the heart of Fenland, had seven oil mills in 1735 and the last closed in 1903. At Gainsborough, where oil milling was first recorded in 1377, extensive multi-storey mills remain along the Trent riverside. Hull was another important centre for oil milling and the first steam-powered oil mill was established here in 1785, although wind- and horse-power remained pre-eminent for some time.

Oil or tallow were essential in soap manufacture, being boiled with soda to produce cakes of soap. Soda was originally derived from the ash of wood or bracken but by the nineteenth century was being produced from salt by the Leblanc process. Soap works were consequently located in areas where both coal and salt were found, such as Merseyside and the north-east. Bristol had previously been a major centre for soap production, particularly of high grade toilet soap, but was affected by government monopolies which gave preference to London in the seventeenth century. Soap was subject to excise duty during the eighteenth century and its production was supervised by Customs officials until its exemption in 1853 at a time of cholera epidemics. The firm of Thomas's in Bristol became one of the nation's largest producers of toilet soap with its 'Puritan' brand. Their factory, with its ornate pan house inspired by the Uffizi Palace in Florence, closed in 1950. One of Bristol's most striking industrial buildings, it was converted to offices and can still be seen on Broad Plain. Other large manufacturers were established in the north-west, notably in Warrington where the firm of Crosfields was established in 1815 to be followed by the Lever Brothers enterprise in 1885. Three years later, Lever moved to a greenfield site on the Wirral to make 'Sunlight' soap and built 800 cottages for his workers, mainly in a half-timbered Tudor style. Port Sunlight remains as an example of the 'garden city' movement which inspired Lever and other public-spirited entrepreneurs.

Animal products

Meat became an increasingly important item of diet in the eighteenth century. Earlier, the lack of winter fodder meant that a proportion of livestock was slaughtered in late autumn and smoked or salted meat consumed until the following summer. One source of fresh meat was doves and pigeons kept in special cotes, the building of which was a jealously guarded feudal privilege during the Middle Ages. These distinctive and often attractive buildings remain a feature of the rural landscape, usually echoing the local vernacular tradition. They are usually square or circular with cupolas often incorporating entrance holes for the birds. By the eighteenth century, they were often demonstrations of architectural pretension on landed estates, even being disguised as classical temples.

During the eighteenth century, greater attention was paid to animal breeding and improved feeds which enabled livestock to be over-wintered. The growing towns and cities could not be fed from their immediate vicinity and the import of cattle into England from Scotland, Wales and Ireland met the demand. Flocks and herds were sold in local markets and then moved on the hoof along recognised drove roads which survive as a landscape feature. These wide green lanes following the lie of the land were once punctuated by resting pastures and inns for the drovers. A whole network of drove roads, marked by limestone walls, can still be traced in the Yorkshire Dales. Further evidence can be found on surviving toll-house tariff boards, which state the charge levied by turnpike trusts for droves of cattle, sheep and pigs. For example, the board on the 1823 octagonal toll-house on the Holyhead Road at Llanfair-pwllgwyngyll in Anglesey states 'For every Drove of Oxen, Cows, or other neat Cattle, per score, the sum of 10d.' Many were *en route* for Smithfield Market in London, which Daniel Defoe described as 'the greatest in the world'. Pigs were shipped from Ireland to Bristol and their road to London passed through Wiltshire. A bacon-curing industry was established here in Calne, Chippenham and Trowbridge which survived long after the droving ceased. Local pig-rearing flourished using surplus whey from the local dairying industry.

Droving reached its peak in the first decades of the nineteenth century, with some 100,000 cattle a year being driven from Scotland alone into England. But cattle lost as much as 6 stones per beast on the journey from Scotland to London and had to be fattened for sale on pastures near the capital. The use, from the 1830s, of steam ships for transporting cattle reduced this weight loss and was particularly beneficial to the Irish trade. Within the next twenty years railways speeded up transport even more and covered pens at both docks and railway yards were built to accommodate the livestock. Smithfield continued to dominate the livestock trade but important provincial markets, with direct rail access, developed. These were often designed to reflect Victorian municipal aspirations and boasted elaborate gates, covered sales rooms, abattoirs and cobbled cattle pens. The Nottingham cattle market in the Meadows area survives as part of the Corporation's East Croft development, complete with a hide and skin market dated 1878.

Such was the demand for meat that imports of foreign livestock increased, but live animals carried the risk of disease. Special markets were established to receive them, such as Deptford Market, close to the Thames, established in 1871. The Woodside Lairage in Birkenhead was opened in 1879 for cattle from the Americas and became one of the country's largest cattle import points, eventually becoming devoted exclusively to Irish live cattle. By 1900, Birkenhead was handling over half a million animals per year. The foreign disease problem was partly overcome by improvements in technology, first chilling and then refrigeration, which enabled the import of carcase meat from the 1870s. Cold stores were built in ports, railway yards and around Smithfield itself. Within less than a century, the age-old practice of droving ceased: London and the major provincial towns were supplied with meat by railway and steamship and the drovers' roads reverted to rural tranquillity.

Fish were more generally eaten in the past than today, and ponds for rearing fish were a feature of monasteries and some country estates, while weirs across rivers were built for fish conservation. The greater proportion of fish was, however, derived from the sea and landed in small harbours around Britain's extensive coastline. The east coast fishermen from Wick to Ramsgate followed the herring shoals which provided the bulk of the fish consumed in Europe. Pilchards and mackerel were caught off the south-west coasts, while oyster beds were a feature of many estuaries. Many small harbours were improved for the fishing fleets during the eighteenth century and tarred wooden lofts were built for net- and sail-making, many of which still survive. A particularly fine group may be found on the foreshore at Hastings in East Sussex. These are weatherboarded structures, two or three storeys high with a pitched roof. The market for fresh fish was confined to the immediate hinterland of the harbours and the majority of the catch was preserved by salting, smoking or pickling. Small fish cellars survive in many of our coastal ports and good examples can be found in the adjoining villages of Port Gaverne and Port Isaac on the north coast of Cornwall. On the east coast, curing houses were usually two-storey buildings with vents below the eaves and in the roof ridge: examples can still be seen in Gorleston and Lowestoft. The industry in Yarmouth operated on a larger scale and an outstanding example is the Tower Curing Works, a tall brick smoke house with an elaborate office block dated 1880. Fish curing was also carried out on the east coast of Scotland and a fine stone-built smoke house remains in Fraserburgh. Containers were essential to the fishing industry and box- and barrel-making works were established: in some cases a cooperage was attached to the curing house.

The keeping quality of fish and other perishable foods could be improved by storage in ice and from the 1820s natural ice, imported from Iceland and Norway, was stored in underground wells. Its use enabled fresh fish to reach a wider market. Only in the 1850s were compression refrigeration ice-making plants coming into use and trawlers and drifters went to sea with cargoes of chopped ice which was used to pack the catch in boxes. The introduction of steam propulsion enabled voyage times to the fishing grounds to be reduced and the introduction of special fast vessels into which catches were transshipped at the grounds enabled them to reach market more speedily. The distribution of fish from port to a wider market was transformed by the steam railway, with special trains travelling overnight to the major towns. Covered fish markets, often with decorative ironwork, are a feature of many town centres, such as the retail market in Leicester which was built in 1877. In London, Billingsgate developed on the Thames side as the major fish market for the capital at the end of the seventeenth century and was served by road, river and later railway. The present building was erected between 1849 and 1853 and enlarged in the 1870s; it was closed in 1982 and subsequently converted to a financial centre.

Many ports grew rapidly as a result of their railway connection. Fresh fish dispatched from Grimsby rose from 450 tons in 1854 to over 165,000 tons by 1902. The Manchester, Sheffield and Lincolnshire Railway, later the Great Central, arrived in Grimsby in 1849 and sponsored the formation of the Grimsby Deep Sea Fishing Company. The first 6-acre fish dock was opened in 1857 and the accommodation for fishing vessels had reached 29 acres by 1900 when 471 steam trawlers and sixty-seven smacks were based at the port. An ice plant, constructed in 1863, replaced imports of ice from Norway and 90 per cent of the catch landed at Grimsby was transported out by railway.[5] Similar growth occurred at Lowestoft through the Great Eastern and at Milford Haven through the Great Western Railway. In Scotland, the West Highland Railway was built to Mallaig purposely to move fish from the specially built harbour and curing houses to Fort William and the main line south. Here, as elsewhere, the fish is now transported by refrigerated lorry direct from the quayside, but the railway remains as a spectacular example of Edwardian engineering.

Salt is an essential commodity with many uses, particularly for the preservation of foodstuffs before canning and refrigeration were perfected. Consequently, taxing it provided a regular source of government income from the medieval period onwards, and it was not until 1825 that the Salt Tax was finally abolished. Other uses included glazes for pottery, soap manufacture, bleaching powder and soda production, together with other chemical processes. Numerous coastal salthouses and salterns were recorded in Domesday as well as some natural brine springs in Cheshire and Worcestershire. Sea water, with its low salt content (3–4 per cent), continued to be evaporated in open salterns such as those at Allonby in Cumbria and Prestonpans in Scotland, which did not close until 1959. Scottish coal masters diversified into salt production as a means of using their unsaleable small coal. The Lymington salterns in Hampshire used small portable wind engines to pump sea water into the concentrating pans. The works closed in 1865 and the windmills had disappeared by the end of the century. Sea salt has been produced since Roman times at Maldon in Essex where the extensive mudflats create high salinity.

Natural brine, with a higher salt content, became the most important source, particularly in Cheshire. It was pumped to the surface, first by windmill or horse-gin, and later by steam, settled in a cistern and then piped to the saltworks. These consisted of salt pans in which the brine was boiled and the resultant salt crystals were shovelled into moulds to form lump salt. The iron pans were covered with lightly constructed timber sheds to allow the heat and steam to escape, since the salt fumes attacked iron and brickwork. William Furnival's saltworks at Wharton, beside the River Weaver, was erected between 1828 and 1832 and illustrated in a series of views which depict the long rows of covered sheds and numerous chimneys sprouting from the boiling houses. The landscape of the salt industry is an ephemeral one, in which the buildings show repeated repairs and reconstruction, nowhere better seen than at the Lion Salt Works at Marston, near Northwich. The Thompson family, who had been in the salt business since 1721, moved to Marston in 1842 and ran their own boats on the adjacent Trent and Mersey Canal. These were the last open pan works in Cheshire, but closed in 1986 and are being retained as a tourist attraction. Rock salt, the source of the brine, was first mined in Cheshire in the late seventeenth century and its production developed on a large scale over the next two centuries. Both the brine and rock salt extraction caused considerable subsidence in the area, marked by numerous water-filled meres and flashes. Buildings were affected to such an extent that under special by-laws in the 1880s, a timber-framed style of construction was adopted for new buildings in Northwich and other salt towns. The attractive Tudor-style buildings of these towns is consequently a direct result of industrial activity, not a late medieval survival. The Brine Subsidence Compensation Act of 1891 introduced a system of levies on saltworks to provide a fund to reimburse property-owners, but the long-term solution lay in controlled pumping, replacing brine with water.[6] Salt is still being mined in Cheshire, much of it for road clearance in the winter months.

Worcestershire, Staffordshire and Teesside also had salt deposits. In the Droitwich area, subsidence was again a problem, but there are few remains of the industry itself. The salt proprietors displayed their wealth in large houses on the outskirts of the salt towns, the most remarkable example being the French-inspired Château Impney at Droitwich. The Cheshire and Teesside landscapes are now dominated by large chemical works, dependent upon local salt for their raw material but requiring coal from elsewhere. Transport has always been of prime importance in the salt industry, since the product had to reach a nationwide market from a limited area. There is considerable map evidence of the salt ways which were used by packhorse trains to carry this much needed commodity into northern and midland England. The salt proprietors of Cheshire played a major role in encouraging the improvement of turnpike roads, but more particularly of the River Weaver which provided an outlet to the Mersey Estuary in the 1730s. Even more important was the Trent and Mersey Canal, opened in 1777, which eventually connected the saltworks to the national canal and river network. The degree of subsidence experienced in nineteenth-century Cheshire can be appreciated today by noticing the elevated position of the canal relative to the surrounding landscape. The history of the Cheshire industry is well told in a small museum in the old workhouse in Northwich.

Milk was not considered so essential in diet in the eighteenth century as it is today, but improvements in farming methods ensured its greater availability towards the end of the century. The Board of Agriculture's reporters recommended purpose-built dairy designs in the first decade of the nineteenth century and many were constructed, especially after the

1840s as part of the model farms which became a feature of Victorian high farming. The availability of spent brewers' grains and other feeds enabled more cattle to be over-wintered and increased milk output. Milk's perishable nature, however, meant that it could not be sent to the growing towns until faster transport was available and so surplus milk in the countryside was made into butter and cheese. Defoe in 1724 had commented on the number of counties which sent cheese to the London market. Fresh milk in London and other large towns was provided from the town cow houses and cows could even be found grazing in St James's Park. In the last two decades of the nineteenth century, one-third of Newcastle's milk supply and one-half of Hull's still came from this source. Building plans indicate that cow houses were even being built at the rear of urban terraces into the first decade of the twentieth century. But town cow houses were breeding grounds for cattle disease, which hit London badly in the 1860s. By now rail transport was available and milk had been brought into London from Essex as early as 1845. The rail-borne milk trade increased rapidly in the hinterland of large towns, but until the introduction of the milk cooler in 1872 farmers could not cool their milk sufficiently for it to be transported long distances. Milk collection depots were established alongside railway lines, particularly in the important pastoral districts west of London. The first was at Semley, near Shaftesbury, opened in 1871 beside the Exeter to Waterloo railway.[7] These depots received milk in churns from the farm and cooled it before dispatch, again in churns, to distribution depots in London and other large towns. The buildings included steam boiler plant for cleaning and sterilisation. Many of the original depots still exist but usually now served by road rather than the railway which created them. Another important area for milk production was north Staffordshire and Derbyshire, from where one-fifth of London's milk was supplied by 1900. The Leek and Manifold Valley Light Railway and the Churnet Valley Railway were both built to serve the dairying industry. A feature of the town distribution depots were the stables, often with upper storeys accessible by ramps, for the horses which drew milk carts round the streets.

Butter and cheese had long been made from surplus milk on the farms and cheese chambers were, for example, a feature of many eighteenth-century farmsteads in the Vale of Belvoir. The rapid distribution of liquid milk by rail destroyed farm-based butter- and cheese-making and both these became factory industries by the 1870s. In Derbyshire, the local agricultural society founded the first experimental cheese factories in Derby and on the Longford estate. Country creameries processed milk from remote farms into butter and, in the second half of the nineteenth century, into evaporated and condensed milk for canning. Some of the earliest creameries can be found in the west of England, where redundant cloth mills were occasionally utilised for this purpose, for example, by the Anglo-Swiss Company, later Nestlés, in Chippenham and Staverton. In Yeovil, the dairy business of Aplin and Barrett founded a creamery and their trade name of St Ivel has become a household word. By-products, in the form of skimmed milk and whey, provided feed for the local pig-rearing industry. As with milk collection depots, steam raising plant was required for sterilisation and the sites were marked by chimney stacks and boiler houses – a far cry from the rural image of the milkmaid portrayed on 'St Ivel' products.

A new product to appear on street stalls around London in the 1850s was ice cream, previously a rich man's delicacy. Its preparation entailed the use of ice which was imported from Norway and also distributed to butchers and fishmongers. The former ice warehouse of the Gatti family at King's Cross stored ice brought along the Regent's Canal from the East London docks. It was kept in huge underground pits, two of which form the centrepiece of a new museum devoted to the commercial and social history of the canal.

Malting, brewing and distilling

Before about 1850, piped water was a rarity and water supplies generally were often polluted. As a result a number of drinks were established as part of the British diet and these varied according to class and region. Beer and porter were the most common drinks consumed by all classes, although the upper classes also drank imported wine. Certain drinks had regional importance, notably cider in the south and west of England and whisky in Scotland. Eventually the national transport network blurred these regional differences as all types of drink became generally available. In addition, Britain's overseas colonies came to supply entirely different kinds of beverages such as tea, chocolate and coffee.

The raw materials for beer and porter were barley, hops, yeast and water, but malt, the product of fermentation of barley grains, could either be used for beer or the liquor further concentrated by distillation to make whisky. While the water from the gypsum beds at Burton-upon-Trent makes good beer, the clear peat-flavoured streams of the Highlands are eminently suitable for whisky. Malting was a seasonal occupation since the germination of barley could be hindered by high temperatures and many maltings closed from April to October. This pattern of work integrated into rural life, since labour was needed in the maltings after the corn harvest was over and ceased before hay-making.

Many villages had a malt-house during the eighteenth century, which supplied malt to local publicans, estate and domestic brewers. The process of turning barley into malt created a distinctively functional building. To promote germination, the barley was first steeped in water and then laid out on the malting floor not more than 8 inches deep. The height of each storey had only to be that of the man who turned the barley and so maltings are still recognisable by their long low storeys with small louvred or centrally-pivoted windows to control the temperature. At one end of the building were one or more pyramidal topped kilns into which the germinated barley was placed for drying. The village malt-house, probably more than any other industrial structure apart from the water-mill, reflected the vernacular building tradition. Few were more than two storeys high and employed local building materials. At Tongham, in Surrey, local chalk and greensand were used, and at West Wycombe in Buckinghamshire the single-storey malt-house in the main street was built of flint. At Stowmarket in Suffolk, at the head of the Gipping Navigation, is a very attractive maltings, built partly of flint and partly of local brick with steeply pitched roofs on both the malt-houses and the kilns. At Burwell, in Cambridgeshire, the attractive stone-built malt-house has a thatched roof with an unusual circular kiln at one end. Local brick was also used in various East Midland locations such as Stathern in Leicestershire, Laneham alongside the River Trent and Southwell, both in Nottinghamshire.

Changes in the brewing industry during the nineteenth century led to the disappearance of the small retail and wholesale brewers who had provided the market for the scattered maltings. Previously a seasonal occupation for the winter months, improvements in technique and the demand for lighter ales meant that it became a year-round industry by the end of the century. The large town brewers who came to dominate the industry constructed new maltings, often at the edges of the brewing towns where there were no space constraints. These new maltings were multi-storey and had several working floors as well as grain stores, kilns and malt garners. They were also built by both brewers and specialist maltsters in the barley-growing areas of the eastern counties (Plate 7). There were transport cost savings to be effected by carrying malt rather than raw barley to the breweries, and waste products could be disposed of to the local farmers. London brewers established large maltings beside railways at Long Melford in Suffolk, and Stansted Abbots and Sawbridgeworth in Hertfordshire, all of which have now been converted into residential accommodation. At Beccles, in Suffolk, a long range of maltings shows several phases of development with kilns interspersed along it between the malting floors, like those at Mistley in Essex. Both were independently owned, like Snape maltings with its numerous kilns, which is now better known as a concert centre. Newark-on-Trent became one of the most important malting centres because of its position beside the navigable river and the later railway network, and vast maltings bear witness to its former pre-eminence.

Bass, the Burton brewers, had an extensive range of maltings in Lincolnshire at Sleaford built between 1901 and 1906. The 1,000 feet long range consisted of eight malting blocks set out in two rows of four on each side of a central building which contained a steam engine and water storage tanks. Railways provided the link to the company's brewery in Burton-upon-Trent, where an extensive railway system once wove its way through

Plate 7 The now derelict maltings on the west of the New Cut at Ipswich. The new channel was created when the River Orwell was enclosed to form the Wet Dock. The range of kilns fronts a series of four- and five-storey floor maltings. The complex, like many other large maltings in East Anglia, lends itself to residential conversion.

the town streets serving the brewing industry.[8] Here, too, several large maltings were built, although malting has now ceased in Burton. Several remain in alternative use, including the Anderstaff, Crown and Clarence Street buildings. The latter, dating from 1883, are unusual in having a large octagonal kiln surmounted by a goat-shaped wind vane on the vent.

While floor malting continued well into the twentieth century, a few maltsters adopted the pneumatic or 'drum' and box malting systems introduced towards the end of the nineteenth century. In these, germination was accelerated by means of aeration of the wet barley and by strict temperature control. The extensive areas necessary for floor malting were no longer needed, but the characteristic kilns were still required. The power consumption in drum and box maltings was high and so traditional floor maltings lingered on in many places. Purpose-built drum maltings at Sawbridgeworth in Hertfordshire and the Plough maltings in Burton-upon-Trent were both built at the very end of the nineteenth century, but only the kilns now betray their former usage. The malted grains needed to be ground before brewing, which was carried out in a malt mill, usually at the brewery itself. This mill was often horse-powered in the eighteenth century but adapted to steam thereafter.

Another essential ingredient for bitter beer was the hop which was first imported from the Continent around 1400 and then grown in England from the sixteenth century. The major hop-growing counties were Kent, Sussex, Hampshire, Hereford and Worcestershire. The hop fields have always had a characteristic appearance with the hop bines trained up chestnut poles, some 4,000 to the acre; only in the late nineteenth century were permanent wirework support systems introduced. Hop-picking was always done by hand until the 1930s and was a traditional summer holiday occupation for Londoners. The green hops were collected in baskets or blankets and taken for drying. This was carried out in kilns, which were normally part of the farm complex rather than urban structures like maltings. They were known as 'oasts' in the south-east and as 'kells' in the West Midlands.[9] The kilns were originally square in section with conical or pyramidal roofs surmounted by ventilation cowls which could be turned into the wind by means of vanes; circular section kilns were introduced in the south-east early in the nineteenth century (Plate 8). Slate-roofed examples of the former survive at Eardisley in Hereford and Worcester, while circular section kilns are scattered over the southern counties of England where they have frequently been converted for residential use. Small farms often have one or two kilns attached to them, as at Crowhurst in Surrey, while clusters of kilns can be seen on larger farms, as at Eardiston near Tenbury Wells. As with maltings, brewers invested in hop-growing and built large kiln complexes like that belonging to Whitbreads at Beltring in Kent. Hops, which were subject to excise duty from 1710 to 1862, were marketed through special exchanges and warehouses. The most impressive example is the London Hop Exchange in Southwark, with its large dealing floor and three galleries of offices above. Worcester still has its exchange and several warehouses belonging to hop factors, including one with a splendid pediment depicting hop-picking. By the end of the nineteenth century, nearly half the hops used for brewing were imported.

Before 1800, domestic brewing was predominant, carried out in the home, on the farm or landed estate. Some publicans brewed for their own sales, while others were supplied by the common brewer who came to dominate the trade by the end of the nineteenth century. Ports like Liverpool, Hull and King's Lynn developed as brewing centres because of the ease of transporting beer by water, but London dominated the coastwise trade as well as serving a vast urban market. The growth of the canal network enabled other towns like Burton-upon-Trent, with its excellent water supply, to become important. Export trades to the Baltic and later to India sustained rapid growth, while changes in taste, particularly the demand for lighter beers, led to the large London companies constructing new breweries in Burton between 1850 and 1875.

The characteristic form of a brewery was dictated by the nature of the process. It made use of gravity to transfer the liquid or wort from one vessel to the next, starting with the mash tun near the top of the building

Plate 8 A farm oast at Chiddingstone in Kent. The original stowage is the central rectangular block with its integral kiln. Production was increased by the addition of four more roundels located along two walls. All the kilns are complete with cowls.

and descending via the copper where the wort was boiled with hops to the fermenting vessel at the lowest level. From this the beer was drawn off into casks. Estate breweries were usually two-storey structures, with louvred windows for ventilation, while a mezzanine floor supported the copper. Good examples can be seen on the National Trust properties at Shugborough in Staffordshire and Calke Abbey in Derbyshire, which provided beer for the large estate workforce. Small breweries can often be recognised at the rear of public houses, like the stone-built example with a vented roof to the rear of the Bell Inn at Seend in Wiltshire, while at Southwick in Hampshire there is still a fully-equipped brewery belonging to the Golden Lion Inn.

The tower brewery reflects a larger scale of production, with a malt store and mill at the top and several floors for duplicate sets of vessels, as is still demonstrated at the Stamford Brewery Museum in Lincolnshire. Steam engines were used to pump water to the top of the buildings and to power machinery. The characteristic profile of a tower brewery must have dominated many townscapes in the nineteenth century and their façades were highlighted by decorative features. Harvey's brewery in Lewes has a half-timbered appearance befitting its Sussex location. In Nottinghamshire, the quality of local brick and terracotta is well shown at the Home Brewery in Basford and the Castle Brewery in Newark-on-Trent, while the Yorkshire town of Tadcaster is dominated by the two breweries of Samuel and John Smith, the latter being a richly ornamented stone building (Plate 9). Perhaps the most elaborate survivor is the Anglo-Bavarian Brewery in Shepton Mallet, Somerset, built in 1872 of local stone with a pediment, balustraded parapet and a variety of fenestration.

The advent of steam, and later electricity, meant that the wort could be pumped between vessels laid out on the same level: the tower construction then became redundant and brewing was carried out in large, rectangular brick-built structures whose functional appearance was often, as in Burton-upon-Trent, relieved by blind arcading or other decoration. The union system of fermentation was developed in Burton, in which the fermenting process was completed in large ranges of barrels, sometimes numbering over 2,000 150 gallon casks in a single system. These had to be accommodated in large single-storey buildings, often around a courtyard which also included cooperages, bottling stores, stables for dray horses and waggon sheds. A water-tower was often the most prominent feature, such as the keep-like structure at Bass No. 1 brewery in Burton-upon-Trent. The brewery became a large complex, often including buildings of various dates. The Star Brewery in Radford, Nottingham, begun in 1852, and Eldridge Pope & Company's brewery in Dorchester are good examples of the fully-developed complex. Some architect-designed nineteenth-century breweries depart completely from the functional tradition, such as Allsop's

Plate 9 John Smith's Brewery at Tadcaster in North Yorkshire was founded in 1758. These fine buildings, of local stone, date from 1884. The tower brewery on the right has an ornate roof and a full-length lucam covers the hoist to the upper floors.

New Brewery in Burton-on-Trent. Now part of Ind Coope, it was constructed in 1859–60 and is fronted by an elegant office block with semi-circular headed windows linked by blue-brick banding.

Whisky distilling in Scotland has a long history and was certainly well established by the end of the sixteenth century. The small traditional distillery had a number of distinctive buildings, including floor maltings and kiln, a large building for the mash tuns and fermenting vessels and the still room where repeated distillation took place in pot stills made of copper. Because whisky has to mature for several years in the cask, particularly the single malts, and did not pay excise duty until it was sold, distilleries had to provide large bonded warehouses where the spirit could be kept under lock and key. A flock of geese was occasionally kept for additional protection! The blending of whisky, using grain spirit and a number of single malts, was not necessarily done at the distillery, but from 1860 was carried out, along with bottling, in bonded warehouses often in ports. Massive examples still remain along the quayside in the port of Edinburgh at Leith, which is lined with grim stone-built warehouses.

The initial malting processes in distilling were similar to brewing, but here, too, floor maltings are now almost completely replaced by the pneumatic or drum process. For the initial drying of the germinated grains, anthracite was used as fuel followed by peat to impart the required flavour. Malt-houses existed alongside many Scottish distilleries, but the malting process, as in brewing, is now generally carried out by specialist maltsters in the east of Scotland.

The Scotch whisky industry has become increasingly concentrated into large production units centralised in the barley-growing area of eastern Scotland. However, many small traditional distilleries remain elsewhere from the nineteenth century, a few operational, others surviving as tourist attractions and some adapted for other purposes. An attractive white-washed group of buildings can be seen in the village of Dalwhinnie in Inverness-shire which was built in 1897–8 by the Strathspey Distillery Co. Ltd but rebuilt following a fire in 1938. Near Pitlochry is the much-visited Edradour Distillery, built in 1837 by the Duke of Atholl. It is the last survivor of the once widespread farm distilleries in Perthshire. On the Isle of Skye at Carbost, the Talisker Distillery of 1832 produces the single malt whisky characteristic of the Western Isles.

In the south and west of England, the traditional drink produced was cider from locally-grown apples. Farm labourers received part of their wages in cider, particularly at the busy times of hay-making and harvest, until the Truck Acts were applied to agriculture in 1878. Cider then became available for retail sale, but remained largely a farm-based craft although some small factories were established in areas close to a large urban market, such as Herefordshire. The apples first had to be crushed to a pulp and this was originally carried out in a horse-driven mill, which consisted of an edge runner in a circular stone trough. These are still familiar sights in cider-making areas, although often now used for decorative purposes. The pulp was then placed in hessian bags and the juice squeezed out in a hand-operated press, examples of which may also still be seen on many farms. A good example remains in the care of the National Trust at Cotehele Mill in Cornwall. The apple juice was then fermented in casks, supplied by local cooperages. The industry is now almost entirely factory-based but still located in the traditional cider-making regions.

Colonial imports

From the sixteenth century onwards, Britain became a colonial power able to import tropical produce which introduced variety into diet. She competed with Holland and France for the long-distance shipping trade, but the colonial trade was protected by the Navigation Acts, first introduced under Cromwell. These restricted both imports and exports to British shipping and although these Acts had to be modified in times of war, they were not abolished until 1849. One of the consequences of the Acts was to encourage a substantial re-export trade in tropical produce, much of which was processed at the British port of entry. The complications caused by collection of customs duties at ports with inadequate facilities encouraged Walpole in the early eighteenth century to attempt the introduction of excise duties on various categories of imported goods, notably coffee, tea, chocolate, sugar, wine and tobacco. He failed, as people feared the introduction of general excise duties (like VAT), and it was not until 1786 that the bonded warehouse system for imported goods was first established. Goods destined for re-export could be dispatched with the minimum of bureaucracy, while the proportion of this tropical produce destined for the home market was taxed by the excise department. In 1800, something like four-fifths of tobacco and coffee imports were re-exported and one-sixth of sugar production. The major ports to benefit were, of course, London and those on the western coasts, particularly Bristol, Liverpool, Whitehaven and Glasgow. Large warehouses sprang up at the ports to house the produce before processing or re-export. Many of these were bonded warehouses in which goods liable to either customs or excise could be stored until removed for sale when duty became payable. These bonded warehouses were characterised by their small barred windows which were a deterrent to theft. Such were the demands upon the Customs and Excise service that warehouses were built in inland towns, the first in Manchester in 1844, followed by Birmingham, Leeds, Sheffield and Bradford. By 1875 there were some 2,400 Customs warehouses and nearly 1,000 Excise warehouses.[10].

Britain has always had an extensive wine trade with France, particularly Gascony, and in sherry and port from the Iberian peninsula. Stored in cool cellars, from 1786 wines were subject to bonding before payment of excise duty after decanting into bottles. At Bristol the Pennant sandstone vaults of Harveys, established in 1796, now house a museum of wine. In London's Docks, there were some 22 acres of wine and spirit storage vaults beneath the quays, usually supported on dressed stone columns. In the St Katharine's Dock of 1827–8 the stone was replaced by iron pillars similar to those on the quayside.

Honey had been used as a means of sweetening food, from at least the Middle Ages, but the acquisition of colonies meant that Britain had a source of cane sugar. The import of this was part of the Slave Triangle between Britain, Africa and the West Indies, with British manufactured goods financing the import of slaves from Africa to the Caribbean. The cane was partly refined on the plantations in the colonies, utilising crushing rollers and other plant manufactured in Britain, and imported as molasses in barrels. Sugar refineries were established in London and in the west coast ports, such as Bristol and Liverpool, there were large multi-storey buildings which contained the plant for the refinement and crystallisation of sugar. The massive Finzel refinery in Bristol, opened in 1836, was one of the largest in the country, but closed in 1881 and its five-storey pedimented buildings have been demolished. Equally, Tate and Lyle's Love Lane refinery in Liverpool was established in 1872, closed in 1981 and subsequently demolished. While the manufacture of sugar from beet was well established in Europe by the nineteenth century, the preference for colonial imports prevented its widespread introduction in Britain until the end of the century when some European sugar beet was imported. The shipping blockades during the First World War caused the rapid introduction of home sugar beet-growing and the construction of the large factories which dominate the flat expanses of eastern England. The sugar-refining process therefore moved from west to east as the sources of raw material changed, and there is little remaining landscape evidence of the former cane sugar refineries which brought such wealth to Bristol and Liverpool.

The cocoa bean originated in South America and chocolate became popular as a drink in the second half of the seventeenth century: the first chocolate house was opened in London in 1657. It was rather an oily drink, but the removal of the cocoa butter by means of the Van Houten press after 1828 made it more palatable. Its consumption was greatly boosted in the nineteenth century by the Temperance Movement, which sought to wean the poor away from gin and beer. Three renowned Quaker manufacturers, Rowntree, Cadbury and Fry, in York, Birmingham and Bristol respectively, promoted its manufacture in the 1860s. The two former both created new industrial settlements with factories and model housing, as can still be seen at New Earswick and Bournville. Fry's city-centre factories were demolished in the 1960s, but their early machinery is preserved at the Blaise Castle Museum in Bristol. The three firms also manufactured chocolate and other confectionery, using sugar.

Coffee, like chocolate, was first introduced as a drink in the post-Restoration period and coffee houses fulfilled an important social and business function in London and major provincial towns. The British captured the market from the Dutch after the Napoleonic Wars and coffee, like chocolate, was subject to duty and the bonded warehouse system.

Tea drinking became popular in the course of the eighteenth century and by the end of the century it was becoming the staple drink of the poor. During the following century consumption rose fourfold. The East India Company imported tea from China and it was not until the nineteenth century that large tea plantations were established in India and Ceylon. The Company built large numbers of private warehouses in the City of London, eventually centralising their operations in a complex around Cutler Street. They later extended their trade to the Pool of London, with warehouses at Wapping, Colonial, Butlers, Milwall, Hays, Tooley Street and Concordia Wharves. Tea imports were a source of tax revenue for the government until 1964 and were consequently stored in bonded warehouses where the processes of bulking and bunging of tea chests for sampling took place. Until 1884 every chest of India or Ceylon tea was turned out on the floor of the warehouse so that the customs could establish net weight and the tea could be mixed using wooden shovels to a uniform quality and appearance. Later only a 10 per cent sample was so treated. Many of the warehouses in the Port of London have been converted to residential and other uses. Purpose-built tea warehouses may also be found outside London, such as the elegant Bush warehouse in Bristol. This was begun in 1835 to handle tea, but now serves as an office and arts complex. In Liverpool, at Stanley Dock, a six-storey brick warehouse of the 1880s also remains.

Tobacco was first imported into Britain in the sixteenth century and achieved popularity as pipe tobacco during the seventeenth, when it was extensively smoked in London's coffee and chocolate houses. Snuff-taking became popular during the Regency period, followed by cigars, but cigarettes were only introduced after the Crimean War. Cigarette smoking grew enormously in the last three decades of the nineteenth century and many large bonded warehouses date from this period. By 1670 half of Bristol's shipping was engaged in the tobacco trade, which was confined to London and Bristol until about 1710 when Liverpool, Glasgow and northern ports such as Whitehaven began to challenge their dominance. In London docks, the New Tobacco Warehouse, or 'skin floor' dates from 1811–13 and is an important example of the structural use of cast iron at a period of rapid evolution. Listed Grade I, the single-storey structure, originally covering 210,000 square feet, has widely spaced cast iron columns which bifurcate in a tree-like manner to support elegant queen post timber roofs with a 54-foot span. It has now been reused as the Tobacco Dock shopping complex. At Liverpool's Stanley Dock a massive new tobacco warehouse was opened in 1900, its thirteen storeys constructed in red and blue brick. The surviving three nine-storey tobacco bonds by the New Cut in Bristol date from 1906, 1908 and 1919 and one of them now functions as the County Record Office for Avon.

Snuff, ground-up tobacco, was first produced in water-powered grinding mills of which the only surviving example is in Sheffield. Cigar manufacture remained on a small scale with factories being established in many of the major towns. Cigarettes and pipe tobacco were, on the other hand, produced in large factories such as those established by W.D. & H.O. Wills in Bristol and John Player in Nottingham. Packaging was an important element and paper and board manufacturers, as well as printers, grew up in association with the mass producers.

Two main factors have affected the development of the food-processing industry in the 200 years we are considering: firstly, the need to feed a vastly increased population and, secondly, the accommodation of changes in taste, particularly for foreign imports, among the wealthier consumers. These demands were met, firstly, by changes in the scale of production and, secondly, by new processing methods. The population grew sixfold between 1700 and 1900, while the proportion of those living in towns, and therefore dependent on the agricultural sector for food, rose from under one-third to over two-thirds by 1900. Improvements in transport enabled food to be moved more quickly around the country, and thereby increased the variety of food available as well as its amount. Some staple foods, particularly corn and cattle, were imported in increasing amounts, and so warehouses, cornmills and meat-processing plants became features of port development. The latter also benefited from tropical imports such as sugar, chocolate and coffee, as sugar mills and chocolate factories were set up in their immediate hinterland. The application of excise duties first to imported tobacco and wine and then to whisky and beer intended for export gave rise to a new class of building, the bonded warehouse, which became a feature first of ports and then of distilleries and inland towns.

Most early buildings connected with the food-processing industry were in rural areas not subject to redevelopment and, consequently, more evidence has survived in the modern landscape of this than of any other industry. Rural water-mills, often in what are now regarded as idyllic riverside settings, have provided attractive residences and restaurants, while the basic simplicity of their machinery has made them attractive to visitors. More restored, and often working, water-mills remain than any other class of early industrial buildings. Many windmills, too, have been restored or maintained as landscape features, and like the water-mill their financial viability depends as much on visitor income as on the sale of their specialised produce. Maltings and even breweries have found ready use for conversion to residential or office use, while the characteristics of maltings, in particular, are reproduced in many new housing and retail developments, ensuring the survival of their particular idiom in the landscape. The buildings of the food-processing industry, therefore, reveal both change and continuity in the industry, with the small-scale producer often managing to stay in business into the twentieth century, while new and more efficient processing plant was built in a different location, often at ports of entry, for overseas imports of both staple foods and luxury goods. Many industrial landscapes reveal this element of continuity to a much greater extent than contemporary documents do, which is why their study is so important in order to gain a balanced picture of the past.

3

Fuel and power for industry

Warmth was as vital to man's needs as food, particularly in the British climate. Fire was first used for heating and cooking, but its ability to transform clay into pottery and ore into metals was soon discovered. Fuel, whether wood, furze, peat or later coal, became an essential element in the economy of developing societies. Likewise, man's ingenuity led him to supplement his own muscle power by making use of other sources of natural power, firstly animal, then water and wind. These met his needs until the great upsurge in the economy in the eighteenth century, when the familiar coal was used to produce a new source of power in the steam engine. The benefit of this invention was nowhere more felt than in the coal industry itself, since the vast expansion of manufacturing industry and the needs of the burgeoning population created massive demand. In the course of the nineteenth century, coal was transformed into a more convenient form – gas – and was used for heating, lighting and to fuel the internal combustion engine. The latter could also be driven by an entirely new form of fuel, mineral oil, the supply of which was limited in Britain until recent times.

Wood, charcoal and peat

Wood was the most obvious source of domestic fuel, but became increasingly scarce as more land was taken into cultivation in the later Middle Ages. Alternative sources were sought, such as furze on the sandy commons of southern England. The enclosure of these during the eighteenth century was a devastating blow to the poor in this densely populated area, far from the coalfields of midland and northern England. In other areas, peat was widely used for fuel and rights of turbary were jealously guarded.

Woodland had long been recognised as an important resource and carefully managed to provide both timber for structural purposes and cord wood for fuel. Areas of woodland were fenced to prevent the intrusion of domestic animals and the trees felled to leave boles around which the coppice poles would regenerate. A number of large trees were left to grow on as standards, and the process of coppicing was frequently enforced at law. It took an average of fourteen years for coppiced woodland to become sufficiently mature for industrial use, although the period was longer in Scotland. To ensure a continuous supply of coppiced wood, therefore, a considerable acreage was needed. The iron-making blast furnace was one of the chief consumers of coppiced wood in the form of charcoal and iron masters often managed their own woodlands. In the Forest of Dean, for example, over 8,000 acres of woodland were needed to keep one furnace going in full production in the seventeenth century.[1] The Dean Reafforestation Act of 1667 tried to enforce woodland management and offenders were tried in the Verderer's Court held in the Speech House; the courtroom is still preserved within the hotel in the centre of the forest. Similar laws were enforced in the Lake District, where a sixteen-year coppice cycle was normal and farmers could often make a better living from wood sales than by grazing sheep. The practised eye can still detect the characteristic form of coppiced woodland.

Charcoal was in great demand for both iron furnaces and forges until well into the eighteenth century. It was made by the controlled burning of wood with a restricted air supply, driving off water and other volatiles to leave a high carbon fuel which had about twice the heat potential of the original wood. Coppice poles were stacked around a central flue on a prepared base and covered with turves to exclude the air. The heap was fired by dropping lighted charcoal into the flue and left to burn for several days. The charcoal burners or woodcolliers managed several heaps at once, living in temporary huts in the woodlands so that they were available at all times to erect screens to protect the heaps from excessive wind and dampen them down as required. The industry was both seasonal and itinerant, but large numbers of charcoal burners could be found in the forests of Sussex, Dean, Wyre and the Lake District. The only evidence which still exists are the numerous charcoal pitsteads which can be detected, particularly in the woods near Lake Coniston. The earthed-up heap for charcoal manufacture continued in use into the twentieth century, although portable metal kilns were introduced in some places. By the end of the nineteenth century, attempts were being made to conserve the volatile products of charcoal burning such as wood oils and alcohol, tar and acetic acid. Large-scale wood distillation works were established, such as those in the Forest of Dean, near Crinan in Scotland and at Blackpill near Swansea, but there is now little evidence of their existence.

White coal or chopwood was kiln-dried wood which was once used in the ore hearth process of smelting lead ores. Circular pits were dug in sloping ground and lined with stone, while a flue admitted air from the downslope side to a fire set in a recess in the pit base. The wood was stacked on bars of wood or stone before burning. Examples may be found near ore hearth sites in Derbyshire and next to the Hoggett Gill smelt mill near Ullswater.

A very important use of charcoal was in the making of gunpowder. Alder buckthorn was the preferred wood and carbonisation took place in sealed metal retorts to prevent contamination. Charcoal, sulphur and saltpetre, the latter often derived from bird droppings collected from dovecotes, were intimately mixed or incorporated under edge runners. The typical incorporating mill was powered by a water-wheel driving two edge runner stones, one on either side of the central wheel. The mill buildings, like others used in gunpowder manufacture, had substantial walls but flimsy wooden roofs. These minimised the damage from accidental explosions, which were fairly common. Subsequent processes were also water-powered and included pressing, corning and glazing the finished powder. Gunpowder works were usually sited well away from human settlement, often in the woodlands where there was adequate water-power. There are remains of several works, for example at Chilworth in Surrey, Faversham in Kent, Powdermills Farm, near Postbridge in Devon (Plate 10) and on the Argyllshire coast of Scotland. Quite often, for safety reasons, the buildings have been deliberately razed to the ground and only the outlines can be traced, for example, along the River Mellte near Pontneddfechan in Glamorganshire, where the gunpowder works of 1857 were immediately demolished following closure in 1931.

The peat blanket which covers much of the highlands and islands of Scotland, the moorland of England and the fenland of East Anglia can be up to 20 feet thick. It has always been an important source of domestic fuel and its extraction in Norfolk from late medieval times has created the landscape of the Broads. Elsewhere, the scars of diggings from which the wet peat is cut can still be seen. The large stacks of drying fuel outside highland crofts illustrate the relative low efficiency of the fuel: 2 lbs of dried peat being equivalent to about 1 lb of good coal. Where wood was in short supply, peat was also used as an industrial fuel. On Dartmoor, peat was used for smelting tin from the beginning of the thirteenth century when turbary rights were granted to the tinners. It was first converted in kilns into charcoal, producing about one-third of its original weight. Peat was used directly for lead smelting, where lower temperatures sufficed. Because of its bulk, large open-sided drying sheds were built for peat storage close to smelt mills, as at Surrender and Old Gang Mills in Swaledale. Further attempts to exploit this natural resource on Dartmoor began in the 1840s, when it was distilled to produce naphtha for use in candles, mothballs and as a lighting oil; the fibrous by-product was made into paper. There are remains of these works at Princetown, Shipley Bridge

Plate 10 One of a series of incorporating mills at Powdermills Farm near Postbridge in Devon. The edge runner mills, in which the saltpetre, charcoal and sulphur for black powder production were intimately mixed and ground while being kept moist, were usually in pairs, one on either side of a central water-wheel. A leat from the River East Dart brought water which was used successively in a series of mills down the slope.

and also at Rattlebrook Head, where a hydraulic compression plant was constructed in the 1870s to produce peat briquettes and a 5-mile railway built down to Bridestowe; this scheme failed but peat continued to be extracted until the 1950s and the line of the railway can still be traced.[2] On the Isle of Lewis, a similar experimental plant complete with railways, designed to extract paraffin from peat, was erected by Sir James Matheson in 1851.

Coalmining

Coal is one of the most common of all Britain's natural resources, but its continued exploitation into the twentieth century has obliterated much of the field evidence for earlier periods of working. Fortunately, documentary evidence, including collections of engravings such as Thomas Hair's splendid *Sketches of the Coal Mines of Northumberland and Durham*, published in 1839, enable earlier landscapes to be reconstructed. Coal occurs in seams rather than the uncertain veins typical of non-ferrous metal-mining, and consequently one area can be worked for a long period of time; some of the pits recently closed in the west Leicestershire and south Derbyshire coalfield have been working for nearly 150 years. Early mining undoubtedly took place on outcrops, which were followed underground as horizontal or sloping drift mines, a method of mining still practised on a small scale by the Free Miners of the Forest of Dean. Of ancient origin, the Free Miners are still recognised and to qualify they have to be men born within the hundred of St Briavels and to have spent a year and a day in a Dean coalmine. Several working drift mines may still be found in the Bixlade Valley, each having its own haulage engine and screening plant. Where the terrain was not suitable for drifts, coal was won from shallow depths by bell pits. Shafts were sunk to the seam and workings extended radially from the bottom, the coal being raised to the surface by a hand windlass. Extraction continued until the overburden threatened to collapse, after which a new pit was sunk nearby. A bell pit survives in the modern landscape as a circular mound with a central depression. Some can still be seen at Strelley in Nottinghamshire, preserved in pasture, whereas at Lount in Leicestershire and Nostell in West Yorkshire a regular pattern of pits can be seen in woodland. Quick-growing trees were often planted on land despoiled by coal-mining for use as pit props, and the name 'Spring Wood' may indicate land used for this purpose. Bell pits can often be detected in ploughed ground by means of variations in the colour of the soil, and many reappear in modern opencast workings.

Seams at greater depth were extracted by deeper shafts, which recent opencast workings have shown to be in use earlier than previously thought. At Lounge in Leicestershire, timber-lined square sectioned shafts dating to *c.* 1450 have been discovered during opencasting. These gave access to pillar and stall workings, a method of operation designed to leave pillars of coal to support the roof and to enable men to work in 'butty' teams. Where these were not far below the ground surface, their outline may be detected on the ground as a grid pattern, especially in winter snow conditions. In many coalfields this method of working was replaced during the eighteenth century by the long wall system where the whole seam was removed, thus creating more surface subsidence. Large amounts of waste had to be brought up the shaft and dumped on the conical tips which once were a common feature of British coalfields. These often loomed above the miners' houses, but following the disaster when a tip in Aberfan buried a school, many have been removed altogether or landscaped as amenity features.

Coal became a more important fuel from the sixteenth century onwards as wood became scarce. It was used, albeit reluctantly, as a domestic fuel once chimneys were introduced into houses as well as for industrial purposes in lime-burning, malting, baking and glass-making. The consequent increase in demand resulted in a dramatic expansion of the industry, with output rising from 210,000 tons in 1560 to nearly 3 million tons by 1700.[3] But output could not be further increased until solutions to the problems of drainage, ventilation and haulage were found. Soughs, underground drainage tunnels, were in common use for draining mines by 1700.

An early drainage system of this kind was constructed for the Bradshaigh family of Haigh Hall near Wigan between 1653 and 1670, and remained in use until 1929: its southern outfall may still be seen. Where the topography was unsuitable for gravity drainage, hand-powered pumps were the only means available, something which limited the depth of the mine. A slightly more sophisticated mechanism was the rag and chain pump, also hand-powered, which consisted of a wooden barrel in which an endless chain carrying discs of leather lifted water from a sump. Several sections could be placed one above the other in the shaft to achieve a lift of about 150 feet, but many of the seams were at far greater depths.

Horse-power was more commonly used for winding coal, but could also be used for raising water in buckets. These winders took two forms: in the first, the cog and rung gin, the winding drum was located over the mine shaft but this cluttered the pit-head. The more common horse engine was the whim or gin where the winding drum had a vertical axle which was directly turned by the horse harnessed to it, and this could be placed at a distance from the shaft. Paul Sandby's water-colour of *c.* 1786 in the National Museum of Wales shows a large horse gin working to an isolated hillside coalmine. The surviving evidence for the horse-gin is usually a level circular walkway and sometimes the stone foot-bearing in the centre. Horse-gins were more reliable, if less powerful, than early steam engines and were often retained as stand-by capstans for raising pumping rods until well into the twentieth century. Reconstructed examples may be found in museums, such as Wollaton Park in Nottingham, but the level platforms survive at collieries which have not undergone extensive twentieth-century development, as on the Bristol coalfield at Nailsea and Painter's Pit in the Golden Valley.

Water-power was less widely adopted in coalmining than in metalliferous mining, but the water balance engine was used for winding in the South Wales coalfield. In this machine, two buckets were joined by a rope passing over a pulley. A full bucket of water descended the shaft, raising a full bucket of coal. At the bottom of the shaft the water was emptied into a drainage level, while on the surface the emptied coal bucket was filled with water to raise a further lift of coal. The remains of a water balance engine may be found at Cwmbyrgwm near Abersychan in Gwent, intact until recently but now sadly vandalised. The presence of these engines can be detected by the name 'Balance Pit' on nineteenth-century maps.

It was the invention of the steam-powered pumping engine that really enabled the coal industry to expand. Thomas Savery's pump of 1698 was intended to drain coalmines but was not successful as its lift was insufficient. Thomas Newcomen's engine, on the other hand, proved to be their salvation and was applied first to collieries, an engine being installed at a colliery near Dudley in the Black Country in 1712. A working replica of this engine has been constructed near its original site at the Black Country Museum in Dudley. These engines used steam at atmospheric pressure to operate a rocking beam which raised and lowered pump rods in the mine shaft. By 1733, when the Savery patent expired, there were upwards of fifty of these 'atmospheric' engines in existence, a comparatively rapid take-up for an expensive technological innovation. By 1778, shortly before the introduction of the Watt engine, there were at least 140 engines at work on the Newcastle coalfield alone.[4] Incredibly, one of the earliest Newcomen engines still survives, because, like many engines, it was reused several times. This is the engine now preserved in Dartmouth as a memorial to its inventor. It began life at Griff Colliery in Warwickshire in the 1720s before being moved to pump water for the Coventry Canal at Hawkesbury in Warwickshire. An atmospheric engine on its original site, built in 1794–5 for Earl Fitzwilliam's collieries, has been retained at Elsecar, near Barnsley. Elsewhere, though, the previous use of atmospheric engines can be traced from surviving buildings. In this type of engine the boiler was placed on the ground with the cylinder above it and the piston rod from the cylinder operating the rocking beam or bob at a higher level. Consequently, the buildings which housed atmospheric engines were tall and narrow and contained massive timbers to support the cylinder. The wall nearest to the mine shaft, the bob wall, was more substantial than the other walls since it supported the bearing for the bob and contained an aperture through which the outdoor section of the bob protruded. When converted for domestic purposes, the infilled archway, substantial wall, massive beams and tall, narrow shape of the building enable its origins to be detected. An excellent example, built in 1804–5, survives on the west Leicestershire coalfield at Moira and there are others at Pontesbury in Shropshire. The atmospheric engine continued in use on the coalfields well into the nineteenth century, much longer than on metal-mining sites where its extravagant fuel consumption made it a costly installation.

The performance and economy of the Newcomen engine were greatly improved by James Watt in the 1770s, especially by his use of a separate condenser. The 'Cornish' engine incorporated his modifications but utilised higher steam pressures and was widely used for pumping purposes. In these engines the boiler was no longer housed under the cylinder but in a separate house and the condenser was located in a well within the engine house. At Prestongrange in East Lothian, a Cornish-type beam pumping engine of 70 inch diameter, made by Harvey's of Hayle, was installed as late as 1874 to pump water, which it continued to do until 1954 and is now preserved as the centrepiece of a mining museum. Also in Scotland, the main beam and pump rod still protrude from the remains of the engine house built in 1854 at Devon Colliery in the old Clackmannanshire coalfield north of Alloa (Plate 11). Elsewhere, only empty shells of former beam pumping engine houses remain. Scott's Pit at Llansamlet, near Swansea, dates from 1819 and excavations by a local amenity group have revealed the base of a haystack boiler. Most are unpretentious buildings, but on the West Cumberland coalfield, the influence of local landowners resulted in more ornate structures. At Jane Pit, near Workington, Henry Curwen built the castellated engine house in 1843 for a beam pumping engine; the horse capstan base for raising the pit rods from the shaft survives alongside. Curwen's rival, the first Earl of Lonsdale, employed Sydney Smirke in the same year to design Wellington Pit in nearby Whitehaven. The castellated gatehouse, battlemented terrace and candlestick chimney of the boiler house still survive as part of an industrial park.

Watt's improvements to the beam engine also produced smooth rotary motion and his winding engines were first introduced on to coalfields in 1784 (Figure 2). They continued to be used alongside atmospheric engines for both pumping and winding into the early years of the twentieth century. By 1811, Farey counted more than fifty steam whimseys in Nottinghamshire and Derbyshire.[5] Fifty years later, many of the beam winders

Plate 11 The beam engine pumping house at Devon Colliery near Alloa in Scotland. This is built of ashlar stone with double hipped roof. The main cast iron beam and pump rod have been retained from the Cornish pumping engine built in 1865 by Neilson and Co. of Glasgow.

were being replaced, first by direct-acting vertical cylinder engines and then by the faster twin-cylinder horizontal engines which continued in use until the early 1980s. Many winding engines had double drums which enabled them to work two shafts, raising in one and lowering in the other: this system can be detected by the presence of two rope apertures in the engine house wall, one higher than the other. The vertical engines needed a tall house, since the winding drum was mounted above the cylinder. One example dating from 1855 which was located at a nearby colliery has been rebuilt in the North of England Open Air Museum at Beamish in County Durham, together with its original timber headstocks. At Bestwood in Nottinghamshire, a vertical twin-cylinder winding engine of 1873 has survived and is now part of the Leen Valley Country Park. The fine brick-built engine house is three storeys high and stands isolated with its associated headstocks, no longer part of the landscape for which it was constructed.

Some engines were built to both pump and wind, like the Neath Abbey rotative beam engine dating from 1845 at Glyn Pits, near Pontypool in Gwent. This mine was sunk by Capel Hanbury Leigh to provide fuel for his nearby iron and tinplate works. The stone-built engine house is located away from the mine shaft and the pump rods were linked by flat rods to a crank arm on the flywheel shaft. Winding drums were also driven from this shaft but pumping was interrupted when winding stopped and it was clearly

Figure 2 A mid-nineteenth-century pit-head scene on the Staffordshire coalfield, from Tomlinson's *Cyclopaedia of Useful Arts*, vol. II, London, 1852. On the right can be seen a rotative beam winding engine with its old-fashioned haystack boiler which is raising coal from two shafts on the left. Cages have not yet been introduced and there are no other buildings at the pit-head. In the background, two other steam engines may be seen, each of them winding from shafts a considerable distance away.

more efficient to have a separate winding engine. A vertical winder was added in the 1850s and this site is now unique in the British Isles with two early steam engines *in situ* on a former colliery site.[6]

More examples still exist of winders with horizontal cylinders, mostly twin cylinders side by side. The houses survive even where the steam engines have been scrapped, since they were easily adapted for replacement electric winders. They did not need to be sturdy multi-storey structures like beam engine houses and were usually tall single-storey buildings with lancet windows, the height being necessary to contain the massive winding drum. At Caphouse Colliery in West Yorkshire, now a mining museum, an engine of 1876 winding over wooden headstocks may still be visited. Several horizontal steam winders remained in use on the Welsh coalfield until the mines closed, and an 1880 example may be seen on Bertie Shaft at the Lewis Merthyr Colliery at Trehafod, now the Rhondda Heritage Park (Plate 12). An engine dating from 1927 forms the centrepiece of the Cefn Coed Mining Museum at Blaen-nant Colliery in West Glamorgan. These horizontal engines could also be used for pumping, the motion being conveyed to the vertical pump rods in the mine shaft by means of a rocking angle bob. A house for this type of engine survives at Calcutta Pit, near Swannington, which drained several mines on the north-west Leicestershire coalfield.

Ventilation could be achieved by natural convection in shallow mines, but as soon as shaft depths were increased a means of removing choke damp and explosive fire damp had to be found. The simplest method was by the use of two shafts, but the air flow could be improved by the use of a fire basket suspended in the shaft. This was dangerous in mines where fire damp was prevalent and in the course of the eighteenth century, underground furnaces, separated from the main shaft, were installed. The surface evidence for these furnaces took the form of chimney stacks over the upcast shaft, as can be seen in Hair's drawing of Hebburn Colliery in 1844 where the chimney is crowned by a cowl similar to that on an oast house. Very few of these chimneys survive, but one can be found at Painter's Pit in the Golden Valley in Avon, with a horse-gin platform alongside.

Since explosive gases could be introduced into these furnaces, colliery explosions were common until a mechanical means of ventilation could be provided. The first of these was the Guibal fan introduced in 1859. None survive on colliery sites, but the characteristic fan house can still be identified. These were built to house fans as large as 50 feet in diameter and 12 feet in width, the fan being almost completely enclosed within the cylindrical housing except for the vent from the mine shaft and the tapering chimney to the atmosphere. The size of the fan can be appreciated at Duke Pit in Whitehaven, where the 36-feet diameter housing is exposed. Other examples may be seen in South Wales at Ynysgedwyn and Cwmbran Collieries, while a brick-built fan house has been listed at New Hawne Colliery near Halesowen in the West Midlands. There were many later

Plate 12 The twin pit headgears at the Lewis Merthyr Colliery at Trehafod in South Wales. The mine is now opened to the public as part of the Rhondda Heritage Park, to commemorate the mining industry of the Rhondda Valleys which finally ended in 1990. The two shafts shown here were sunk in 1880 and 1890 and were named after the sons of the owner, W.T. Lewis, later Lord Merthyr. On the right-hand Trefor shaft, the twin-cylinder horizontal steam winding engine has been retained. This was the upcast shaft for ventilation; the head of the shaft is enclosed and air extracted from the mine by a fan driven from the engine house on the right.

types of fan which were designed to operate at higher speeds, but surviving examples are few. The Navigation Colliery at Crumlin in South Wales has a splendid collection of red-brick buildings with Pennant sandstone quoins, including two built in 1911 to house Walker 'Indestructible' fans. One of the engines which drove these fans is on display in the Maritime and Industrial Museum in Cardiff. A gaunt red-brick Walker fan house stands by the road at the Wynnstay colliery near Wrexham.[7]

The landscape with which these engine houses would have been associated changed dramatically between the eighteenth and the end of the nineteenth centuries. Paintings and engravings of late eighteenth-century pit-head scenes indicate an atmospheric pumping engine, a steam winding engine, a chimney to the boiler house, simple headstocks, a heap of coal and very little else (see Figure 2). The mines were still operating in good seams and little waste was brought out, so pit heaps were small. A hundred years later, there were far more surface buildings, including those housing screens to separate the lumps of coal from the slack, picking belts and washing plants together with stores and workshops. The Lady Victoria Colliery at Newtongrange, south of Edinburgh, designated as a museum, displays the full range of surface plant found on a late nineteenth-century coalmine and the extent of change can be appreciated by comparison with the small drift mines still working in the Forest of Dean, making use of seventeenth-century methods with few concessions to twentieth-century technology.

Mining settlements

The landscape of coalmining extends beyond the immediate environs of the pit-head. Coal was a heavy and bulky product to distribute, but to begin with was carried in panniers on packhorses or carts to markets. More distant distribution was undertaken by water wherever possible, hence the early development of the north-eastern coalfield with its seaborne trade to London and the east coast. It was for the carriage of coal that most of the pioneering developments in transport took place, namely river navigations, canals, horse-drawn waggon ways and, later, locomotive railways. Thus the coal industry not only provided fuel for steam engines in mills and factories but also created an infrastructure for the distribution of their products. The effect of transport on the landscape will be considered in Chapter 7.

The expansion of the coal industry had a widespread effect on settlement patterns. The miner, often on shift work, had to find accommodation near the mine. At first, while mining was combined with farming and usually seasonal in nature, no special housing was provided and miners built their own squatter dwellings on common land. Arthur Young refers to Mr Danby's colliery on the edge of his moors near Swinton in South Yorkshire, which employed many hands:

> The cottages of the colliers are scattered about at no great distance. Each had at first a small garden, which from the great foresight and refined politicks (for I give the conduct no other name) of their landlord, grew into little farms. . . . Observing some of the men to cultivate their gardens better than their comrades, he made them an offer of inclosing from the moor a field for each, contiguous to their gardens, that they might raise their own corn instead of buying it . . . there is no a collier without his farm, each from three or four to twenty acres of land.[8]

This pattern of scattered housing, with few nucleated settlements, can still be seen in Shropshire and parts of the East Midlands, notably the area around the Beaumont Collieries at Coleorton in Leicestershire and at Cossall in Nottinghamshire, the latter on land once belonging to Beauvale Priory. These houses, like those at Swinton, usually had large plots of land attached to them from which a subsistence living could be earned by miners when laid off work. This combination of mining and smallhold-ing survived until at least 1900, miners in pit villages often cultivating large allotments.

One of the consequences of extensive underground mining was surface subsidence, which affected much of the housing erected near the pit-heads. The Swedish traveller Eric Svedenstierna noticed this on the Shropshire coalfield as early as 1802:

> The method of working is quite simple: first of all the coal is taken out within a certain part of the area and afterwards as much as ironstone as is required, after which the mine is abandoned. Everywhere in this district collapsed shafts and depressions in the earth can be encountered, even where there are houses and gardens. Some houses were in fact broken down, or stood so slantwise, that I could not conceive how any one could live in one of them.[9]

Housing affected by subsidence was a common phenomenon in mining districts, but has disappeared so completely in recent years that the Black Country Museum has seen fit to re-erect an affected house from Gornalwood to remind people of vanished landscape features.

As coalmining became more capital-intensive, a more regular pattern of work was demanded by the coal-owners and purpose-built accommodation became more of a necessity. This could take the form of barracks in which dormitory accommodation was provided for the working week and the men returned home at the weekend. Alternatively, coal-owners provided basic family housing in single- or two-storey blocks: the former were more common in Scotland and the north of England. At Rowanburn, in Dumfries and Galloway, three ranges of single-storey cottages with associated coal stores and wash-houses survive from the late nineteenth century. An early two-storey example was Stone Rows on the Ashby coalfield in Leicestershire, built on wasteland in 1811 by the Earl of Moira to house fifty families. These were built on to the street without front doors, the only access being via a passageway along the back. The contrast with Rawdon Terrace, built in 1864 in the same village, indicates the change in attitude in fifty years. These later houses were provided with small front gardens and long allotments at the rear. On the South Yorkshire coalfield, Earl Fitzwilliam sunk the New Colliery in 1795 at Elsecar, which grew from a scattered hamlet to a considerable village in two generations. Between 1796 and 1801, Station Row and Old Row were built to the design of John Carr, who had worked on Earl Fitzwilliam's great house and stables at Wentworth Woodhouse. The curving terrace of Reform Row was added in 1837 and a Miners' Lodging House especially for young colliers at the recently opened Simon Wood Colliery was built in 1854. The religious and educational needs of the community were met by the provision of Methodist and Anglican churches and schools, while a flour mill of 1842 and private railway station of 1870 also survive in what is now a conservation area.[10]

Mining companies as well as aristocratic entrepreneurs regarded the provision of housing for the miners and their families as their responsibility, accepting it as part of the overall capital costs of mine development. The colliers were tied to their houses and could be evicted from them during strikes. In the East Durham coalfield, colliery villages were often built on greenfield sites adjacent to new mine sinkings. Between 1801 and 1851 the population in this area grew from 3,763 to 42,000, with sixteen new collieries, and the number of houses increased from 771 to 7,876. At Hetton-le-Hole, 100 houses a year were built between 1821 and 1831, and a group of five of these has been moved to the North of England Open Air Museum at Beamish. The Wingate Grange Colliery Company built 461 houses between 1837 and 1841, their costs ranging from £30 for pitmen to £75 for overmen and £600 for the viewer. By 1839 the company had spent £26,000 on surface installations, of which the greater part was for houses.[11]

In West Yorkshire, single- and two-storey miners' houses were built in the late eighteenth century, but after about 1825 only multi-storey dwellings were erected. At New Sharlston, Long Row was built in 1864, the same year as Rawdon Terrace in Moira, but twenty-nine of its houses were built back-to-back, as was common in the county. By 1890 the mining company had built 174 houses along with a chapel, school, penny bank, literary institute, library, coffee house and co-operative store. Most of these colliery houses were monotonous, red-brick terraces, but after 1900 there was more variety in the types of houses constructed. The nineteen Voysey-designed dwellings at Whitwood, with their characteristic broken roof line and rough-cast rendering, were built in 1904–5 and may be compared with the austere rows at New Crofton with their unimaginative names, First to Seventh Street, built at the same time. The Whitwood houses had front and rear gardens but only separate allotments were available at New Crofton. The planned villages of Havercroft and Upton, built in the 1920s, illustrate the initiative in West Yorkshire of the Industrial Housing Association, an independent body set up by the colliery-owners to alleviate the post-war housing shortage.[12]

Similar comparisons can also be found on the Nottinghamshire–Derbyshire coalfield, where the Bolsover Company was a major provider of tied houses. Their earlier development below the castle at Bolsover contrasts sharply with their later village at Forest Town, the former being laid out imaginatively around a central green space, while the latter reverted to a tightly-packed regimented pattern (Figures 3 and 4). Although providing a better standard of housing for the miners, the new houses were criticised by D.H. Lawrence as 'set down like a game of dominoes, with spaces and gardens, a queer game that some weird masters were playing on the surprised earth'. He resented their rawness in the landscape and preferred 'the intimacy and smallness of colliers' dwellings over a hundred years old'. By 1930 the Bolsover Company had built nearly 3,000 houses, some 44,000 people being dependent upon the company for a livelihood.

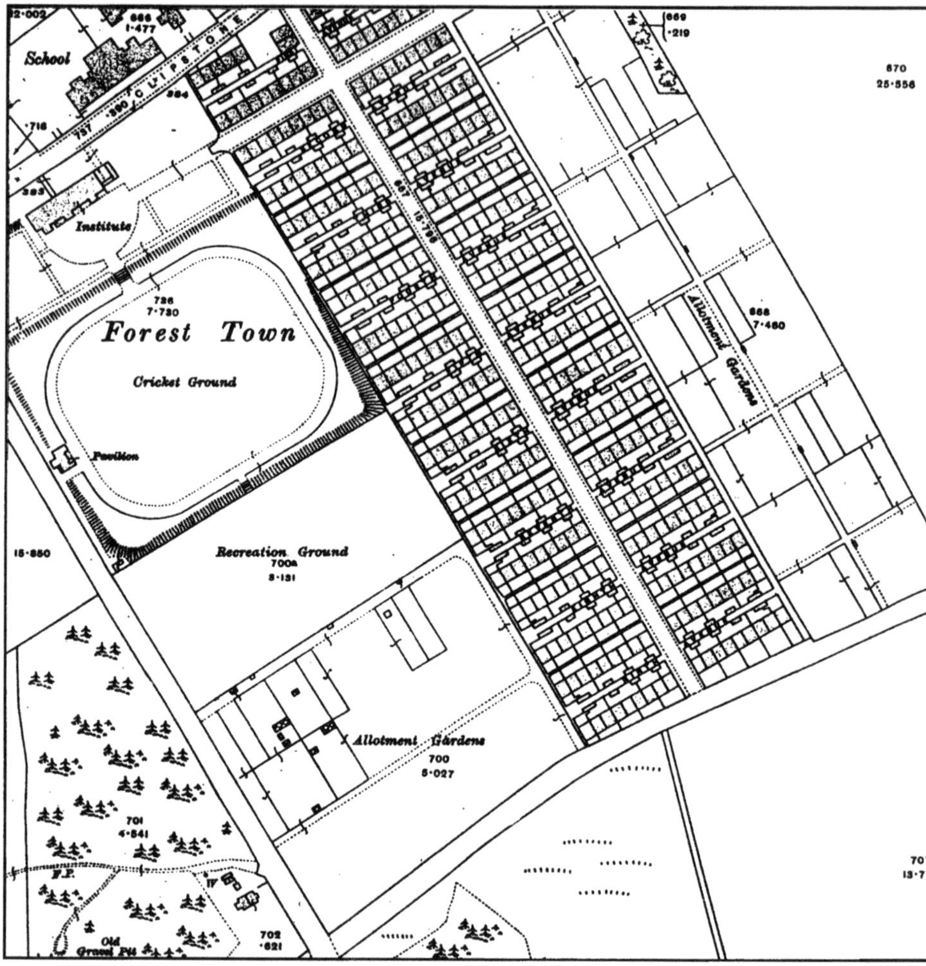

Figure 3 An extract from the 1914 Ordnance Survey map showing the regimented rows of small terraced houses built by the Bolsover Company, named Forest Town, to house their workers at the new Crown Farm Colliery near Mansfield in 1905. The housing density contrasts sharply with that of New Bolsover shown in Figure 4: here there are 324 houses laid out in nine rows of eighteen on each side of a central access road (Nottinghamshire Sheet XXIII No. 14).

Figure 4 (opposite) An extract from the 1914 Ordnance Survey map showing the more imaginative layout at New Bolsover, created for the same colliery company in 1890. There are 194 houses arranged in double rows around three sides of a square: a waggonway ran between the rows along which the coal was brought and nightsoil removed. Semi-detached overseer's houses were built on the access road to the new village (Derbyshire Sheet XXVI No. 6).

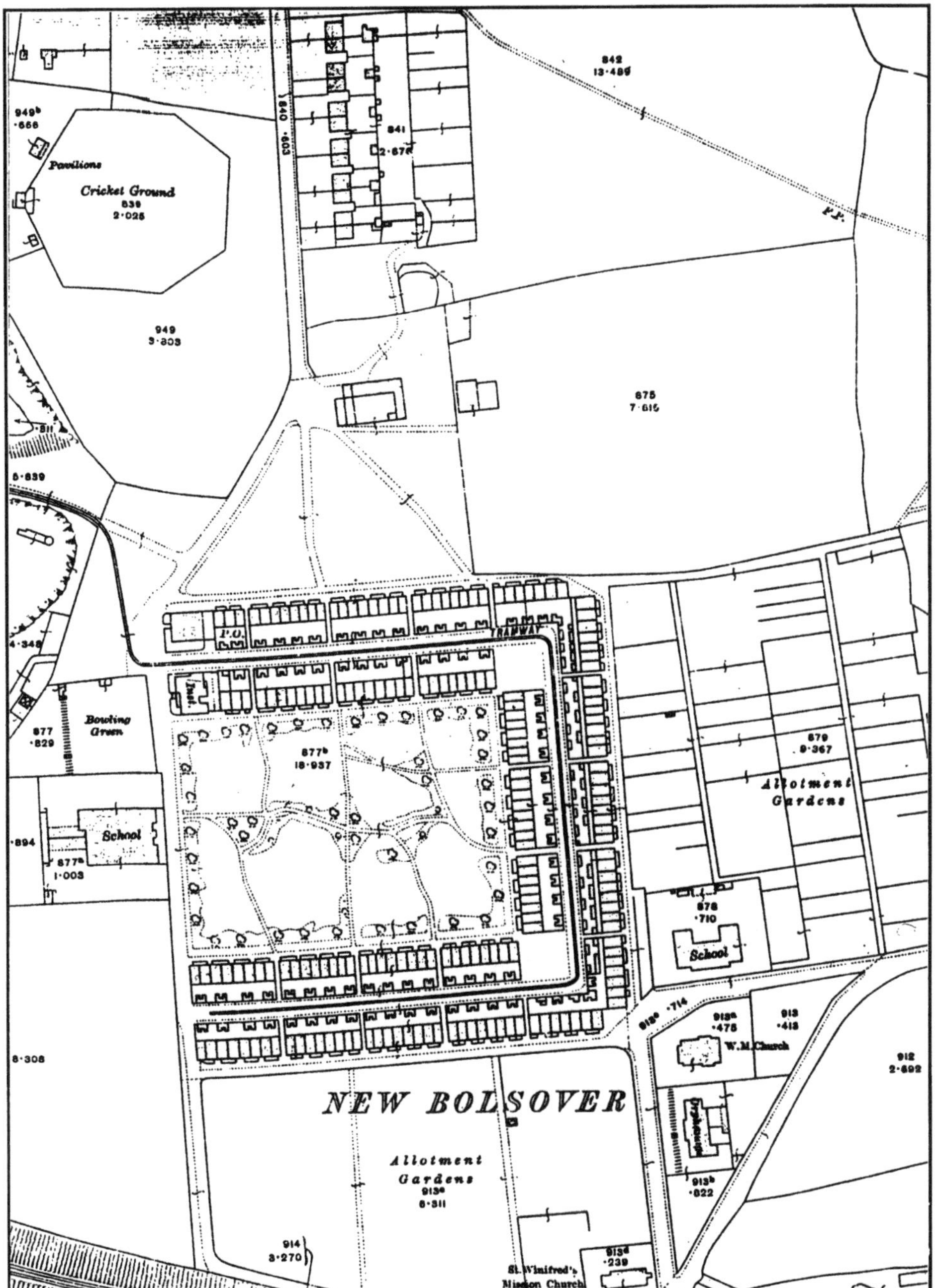
Pavilions
Cricket Ground
Bowling Green
School
P.O.
TRAMWAY
Inst.
Allotment Gardens
School
W.M.Church
NEW BOLSOVER
Allotment Gardens
St. Winifred's Mission Church

On the South Wales coalfield, many of the early mines were owned by ironmasters, and it was not until the 1840s that collieries were specifically opened for coal shipment. Much of this development took place in the steep-sided Rhondda Valleys, which were transformed from a pastoral area of small farms into a vast urban conurbation inhabited by an industrial class new to the area. Whole villages were built as the coalfield extended northward in the pursuit of the excellent bituminous steam and coking coal. The original mine-sinkers were accommodated in wooden barracks which were later replaced by houses, many crudely built because of the short-term leases granted by landowners to the mine operators, who therefore provided only the barest necessities of communal life. Long parallel terraces of houses appeared along the valley bottom and up the sides which were stripped of trees for use as pit props. The 10-mile long Rhondda Fawr Valley floor became an almost continuous built-up area from Porth to Blaen Rhondda. Some of the earliest houses were only two-roomed although four rooms became the norm. Not all were built of the local Pennant sandstone, however, as *The Times* reported in 1867 on Ferndale in the Rhondda Fach Valley: 'almost all the population of 800 is lodged in houses built of wood, like American log huts'. The number of inhabited houses in the Rhondda Valleys increased from 2,710 in 1871 to 26,250 by 1911, 16,000 of which had been built by speculative private landlords. With an average of around six persons per house, overcrowding was greatest between 1871 and 1891 when conditions were as bad or worse than anywhere in Great Britain.[13] The coalfield declined after the First World War and recent surface clearances and safety measures concerning mine waste tips have seen the green landscape restored, but the miners' terraces following the contours of the valley sides may still be seen at Treherbert, Cwm-parc, Treorchy, Pentre, Porth and Cymmer.

Similarly, housing and the rapidly disappearing waste tips are all that remain of the Scottish shale oil industry. Oil-bearing shales were discovered in the Coal Measures of West Lothian and Lanarkshire, where from the 1850s onwards a number of retort plants were established to distil off the oil which was used as a lubricant and a lamp oil. By 1870, some ninety-seven firms were employed in the industry which is now extinct. There are few remains of the retort plants and the distinctive pink heaps or 'bings' of processed shale, with either flat or conical tops, have been cleared for road ballast. Several villages remain where housing was constructed for the workers in the industry, among them Seafield, Tarbrax and Addiewell.

Coke and gas production

Charcoal had been used for smelting iron and other metals for centuries, but was expensive to produce and its fragility limited blast furnace output. Coal, in its natural form, had limitations as a fuel for ironworking because it contaminated the metal by either the addition of too much carbon or the unwanted effects of sulphur and phosphorus. The process of carbonisation to rid coal of these undesirable elements involved heating it in a closed vessel from which air was excluded, producing coke and other products which varied according to the quality of the coal and the temperature of the process. At high temperatures, a purer coke was produced which was more suitable for metallurgical purposes whereas lower temperatures produced both coke and an inflammable gas suitable for domestic use.

Coke was originally produced on a small scale for use in smithing, but it was the switch to the use of coke instead of charcoal for iron smelting in the mid-eighteenth century which encouraged large-scale production. Demand also increased a century later with the introduction of steam traction on public railways. The railway enablement acts stipulated that locomotives should 'consume their own smoke' and coke was used until improved fire-box designs enabled the use of cheaper raw coal in the 1850s. The large-scale production of coke was originally carried out in open heaps about 12 to 15 feet in diameter and 2 feet high or in kilns similar to a limekiln. Beehive kilns were introduced in the mid-eighteenth century and the demands of the metallurgical industry led to large batteries of such kilns being built, particularly where good coking coals were mined as in South Wales, the North Midlands and the north-east. Only in these large batteries of beehive kilns, which could number up to 600, was any attempt made to utilise some of the by-products. The gaseous materials were either ducted under the kilns to heat them or used in waste heat boilers for steam raising. The beehive kiln had several disadvantages: the special brick linings required regular renewal; supplies of suitable low ash coal became exhausted; only some 5 tons of coke were produced in each three- to four-day firing and potential income from by-products was wasted. The last working range of beehive kilns was at Whinfield, near Rowlands Gill in Tyne and Wear, built in 1861 for the Marquis of Bute. Since their closure in 1958, six kilns have been preserved out of the original range of 193 which during its period of peak production produced 68,000 tons of coke per annum.[14] Other more derelict beehive ovens may be seen at Oswaldtwistle in Lancashire, beside the Leeds and Liverpool Canal. In the late nineteenth century, over 1,000 beehive coke ovens were working in Derbyshire to produce metallurgical coke. Their importance in the landscape has been recognised by the scheduling of a double range of forty-eight kilns at Summerley Colliery, between Dronfield and Chesterfield.[15] The first coke ovens where the by-products were regularly collected were introduced at Crook in County Durham in 1882. Each oven in this battery of twenty-five rectangular ovens produced 4.5 tons of coke within a 60 to 72 hours' burning cycle. Loading and unloading was carried out mechanically and the glowing mass of coke was pushed out into iron waggons before being quenched.[16] Working examples may still be seen at Wingerworth, Bolsover and Abercynon. Coke oven gas was also piped for domestic use and around Sheffield a 1,400 mile gas grid was laid down to serve an area of 328 square miles.

For over 100 years before this, gas derived from coal had been used for domestic purposes. William Murdock made the first successful experiments in the production of inflammable gas, using it to light his house at Redruth in 1792. Soho Foundry in Birmingham was lit by gas to celebrate the Peace of Amiens in 1802. In the following forty years, coal gas plants were erected by private companies and municipal authorities to provide gas for lighting; fifteen cities had such plants by 1820, and over the next thirty years most towns and villages of any consequence acquired a piped mains supply. Where possible the works were built near navigable waterways and later alongside railways for convenience of coal supply. In addition, many large private houses and manufactories in isolated areas installed their own private gas-producing plants. The use of gas for cooking did not become common until the 1870s and for space heating until the 1880s. Originally, coal was distilled in iron retorts and the gas produced was passed through water to remove some of the impurities. The iron retorts had a short life and by the 1850s were replaced by horizontal fireclay retorts, followed in the 1880s by the inclined retort and then the vertical retort in the 1890s. Fully mechanised loading and unloading systems had been available since the 1860s. The basic process remained more or less unchanged until the advent of North Sea and natural gas in the 1960s.

The coal gas plant was characterised by a series of distinct but linked sections – a coal storage area, retort house, hydraulic main to prevent blow-back into the retorts, air-cooled condenser to remove the volatile tars, exhauster to draw the gas through, scrubber to remove ammonia and residual tars, purifier to remove sulphur, station meter to measure output, gasholder for storage and a pressure governor to regulate the public supply. Two examples of an almost complete public works remain; the first, at Fakenham in Norfolk, was scheduled as an Ancient Monument after closure in 1965. These works were opened in 1825 and the present horizontal retorts date from 1846, although they have been rebuilt since, and the by-product plant is post-1900. Two gasholders remain; the oldest, with a single lift, erected in 1888 and the other, of one-piece riveted construction, built in 1924. The other example, at Biggar in Scotland, dates from 1839, although the surviving works, which closed in 1973, are largely a rebuild dating from 1914. A battery of nine hand-charged retorts and the other plant are in the care of the Royal Scottish Museum, and give some idea of the scale of a small town gasworks in the mid-nineteenth century. There are several remaining examples of private gas plants on country estates, notably at Holkham in Norfolk, where the louvred retort house can still be seen.

The advent of natural gas has meant that, in most cases, the gasholder is the only feature remaining of the coal-fired works. There were two basic types: the earlier one using a water seal and the later a dry holder where the gas was contained beneath a moving piston. The earliest, and more common, water-sealed holders were inverted iron cylinders with the open end immersed in water, the cylinder being guided by a frame of surrounding uprights. Two and three lift telescopic holders were developed in 1824; as one lift rose it picked up the next, forming a water seal between the two. In the 1880s, the spirally-guided holder was devised which dispensed with the surrounding framework. Many early gasholders were totally enclosed in brick buildings, in theory to guard against explosions, but this practice was later abandoned. An example survives in Warwick where two octagonal blocks with louvred lanterns were built in 1822 to house the gasholders. A fine group of five telescopic holders with lattice cast iron frameworks remains at King's Cross in London, three of them dating from the 1860s onwards (Plate 13). Dry holders were first perfected in Germany and not introduced into Britain until the 1920s. These fixed cylinders, which do not rise or fall, are a permanent feature of the landscape and examples both of a dry holder and the telescopic and spiral variants of the wet holder may be seen at West Bromwich.

Plate 13 Gasholders at King's Cross in London, two still in use on the left and three disused on the right, which were erected from the 1860s onwards for the Imperial Gas Light and Coke Company. The holders on the right are listed and are unusual in that their column structures are interlocking to give additional support. In the foreground are the St Pancras Locks on the Regent's Canal which was cut between 1812 and 1820 from the Grand Junction Canal at Camden Town to the River Thames at Limehouse Dock. In the background may be seen Barlow's 1868 trainshed of St Pancras station and the ornate roof structure and clock tower of the adjoining hotel built five years later.

As well as being burnt for lighting and heating, coal gas could be used as fuel in an internal combustion engine. Unlike a steam engine, where the fuel heats water to provide steam to move a piston in a cylinder, in an internal combustion engine the fuel does the work directly within the engine cylinder itself. These engines provided a portable power source and were used in a variety of contexts, including farms, mine sites, factories and pumping stations. Gas engines were available from the 1860s, but in the 1880s light oils, petrol or benzol, which vaporise at room temperature, provided an additional fuel source. Gas and oil engines were more compact than steam for the equivalent power output and did not require boilers or a continuous water supply. Factories in towns could utilise public gas supplies, but large factories and mills built their own gas plants both for lighting and powering engines. Their widespread use can be identified from the fire insurance plans issued by the Goad Company which were first compiled in the 1880s.

The depletion of natural woodlands by the seventeenth century for shipbuilding and house construction as well as for fuel had become a serious problem. Charcoal, needed in ever-increasing quantities for the iron industry, aggravated the problem which was to a certain extent alleviated by woodland management in the form of coppicing. Britain had ample reserves of coal, but getting these at depth was hampered by problems of drainage, haulage and ventilation in the mines. Finding a solution to these problems encouraged early technological innovation, and the steam engines of Thomas Newcomen, James Watt, Richard Trevithick and others used coal as a fuel to generate steam. So successful were they that coalmines were able to supply the ever-increasing demand, even after the blast furnace was adapted to use coke instead of charcoal as a fuel. An unexpected benefit of coke-making was the production of coal gas, which greatly improved the quality of life in the course of the nineteenth century by providing both lighting and heating in home and factory.

Large areas of underground coal could be exploited from one or two shafts with the aid of mechanised underground haulage, and therefore coalmines, unlike metalliferous mines, remained in one place for many years. New plant replaced old, and the coalmining industry has always tended to destroy its own past. This has been accentuated by the collapse of the industry in the late twentieth century, and closure of mines has been rapidly followed by total clearance of buildings and structures. Reclamation schemes have removed the large waste tips which once disfigured the mining areas, and in some places now only the close-packed rows of miners' houses remain as a testament to a former industrial landscape.

4

Metals in the service of man

The complex geological structure of the British Isles and the consequent variety of rocks and minerals within comparatively small geographical areas has meant that few regions have escaped some form of extractive industry in the past. The upheaval and destruction caused by mining and quarrying have resulted in these landscapes, perhaps more than any other industrial landscapes, being condemned rather than studied and frequently swept away without record. Past extractive industry has created areas of derelict and contaminated land which, in the current climate of environmental concern, is now subject to reclamation schemes. At Foxdale on the Isle of Man, two chimneys point upwards away from a barren landscape, the spoil heaps of a group of twelve mines which provided much of this country's lead and zinc in the nineteenth century. In Cornwall, the parish of St Day was at one time the richest copper-mining district in Great Britain, but the consequent scale of operations has left a vast area of devastation on the Consolidated Mines sett, which was abandoned in 1857. Pevsner, describing the Commissioners' Church of 1828 in St Day, remarks that it 'looks over a landscape of deserted mines, like so many monuments to the passing of human achievement, more clearly moving than the artificial picturesque mementoes in eighteenth century gardens'.

It is possible, however, to view these landscapes as evocative rather than as eyesores left by past exploitation. Artists found inspiration in them, both J.C. Ibbetson and J. 'Warwick' Smith painting water-colours of the vast opencast copper workings of Parys Mountain, while he, Paul Sandby and Joseph Wright of Derby recorded the drama of the iron forge. Tourists walking the Miners' Track to Snowdon may marvel at the bleakness of the barracks where copper-miners lived out their working lives, with visits to their families only from Saturday lunchtime to Sunday night, and then at the expense of a long walk there and back. Elsewhere, isolated communities were established, perhaps most spectacularly at Wanlockhead and Leadhills, high up in the Lowther Hills of southern Scotland, where the single-storey miners' cottages cluster in the latter village around one of the earliest working men's libraries ever to be built, dating from 1756. The National Trust has accepted the picturesque quality of some of the tin-mines of Cornwall and protects not only the working engines at East Pool but the cliff-hugging engine houses at Wheal Coates and Rinsey Head. The work of industrial and mining archaeologists has helped to create public awareness of the history of extractive industries, situated as these so often are in areas undeveloped because of their terrain, but now regarded as the lungs of an urbanised population.

Mining for non-ferrous metals

The surviving landscape evidence is essential to our understanding of the history of non-ferrous metal mining, since documentary evidence is both fragmentary and unreliable. The elaborate equipment advocated in the nineteenth-century technical literature was never adopted in some areas or was modified to suit local conditions, and particular methods were therefore by no means universal. In addition, since many mining concerns were public companies seeking to raise capital, there is the danger that written accounts were exaggerated to impress potential shareholders and buildings said to be under construction were in practice never even begun. Generally, the same technological advances discussed in the last chapter which promoted the growth of coalmining also affected metalliferous mining, but in the case of non-ferrous ores the pace of extraction was slower. The exhaustion of lead, tin and copper veins and new overseas sources meant that, anyway, most development had ceased by 1900, and these abandoned mines present a relict nineteenth-century landscape.

Mineral veins, known also as lodes or rakes, which outcropped at surface had been largely worked out by the eighteenth century, although the open-cuts used to work them are still a feature of the landscape. Many of the Derbyshire rakes, such as Earl Rake, Bradwell, were worked upwards to the surface and form dangerous open trenches. Other rakes were exploited by means of bell pits sunk about 30 to 50 feet into the lode. At surface, these appear as hillocks with a central depression strung out at 10- to 15-yard intervals, resembling the activities of some gigantic mole. Examples can still be found in the northern Pennines and Derbyshire, where the term 'rake' is a common feature on the Ordnance Survey maps. Single bell pits often indicate an unsuccessful trial working. By contrast, in nineteenth-century Cornwall, with tin-mining being carried out to a greater depth than was usual in lead mining, the course of a lode was more usually marked by a line of pumping engine houses; a particularly good example is the Great Flat

Lode, running east–west to the south of Camborne and Redruth in Cornwall. This lode is first marked by the two engine houses of Wheal Uny and then passes below the prominent remains of Basset Mines, comprising both dressing floors and engine houses, to West Wheal Frances, Grenville United and South Condurrow. The shallow valley over the Great Flat Lode has been relatively little built over, so that much of its nineteenth-century mining landscape survives and it is to be conserved as part of the Mineral Tramways Project, designed to encourage visitors to explore inland Cornwall.

Where veins were cut through and exposed by rivers or streams, adits – horizontal or slightly inclined tunnels – were driven in to follow them. The dank, gloomy entrances to these tunnels are a feature along most watercourses in mineralised areas; they can be seen well, for example, close to the waterfalls of Pistyll Rhaedr and Pistyll Cain near the Gwynfynydd gold-mine in Gwynedd, where the water running from them is still stained with minerals.

If Nature had not done it for them, the miners themselves used water as a means of detecting veins by the process known as 'hushing'. This involved collecting water in a dammed-up area and suddenly releasing it to wash away the overburden in the hope of revealing the vein. In this country it seems to have been first practised by the Romans at Dolaucothi in Dyfed, where two aqueducts, one about 7 miles long, brought water from the Afon Cothi and the Afon Annell to two large reservoirs well above the level of the workings, from which it could be released when required [1]. The archaeological remains here are now in the care of the National Trust. Hushing could only be carried out where a steep hillside was to be explored, and the labour involved in bringing water to the crest of the hill was considerable. The debris from a hush fanned out at the foot of the hill to be sorted over by the miners. Hushes can be detected both by this debris and by gashes with exposed rock-faces in the hillside, although both can also be features caused by natural rather than man-made torrents. The remains of dams and sluices at the crest of the hill confirm man's interference. Hushes are particularly common on the slopes of the Pennines, where the word is often used to mark a topographical feature, such as Old Field Hush Gutter near Keld in Swaledale, which was worked from 1738 to 1846. Elsewhere in the Yorkshire Dales, the veins in the sides of the valley of Gunnerside Gill are marked out by large-scale hushes, while on the western slopes of Arkengarthdale, the junction of the Stothert and Blackside Veins is a maze of hushes and opencast workings on a very large scale.

Mineral veins, then, were detected by outcropping or hushing and exploited by shallow shafts or adits. Even by the early eighteenth century, however, most of the more easily accessible veins had been worked out and shafts had to go deeper. As in coal-mining, this led to similar problems of drainage, ventilation and means of access and egress for both men and minerals. Rag and chain pumps were used for raising water while hand windlasses lifted water, men and minerals. In Derbyshire these were often sheltered in small stone buildings known as 'coes', and where these survive they may often have a dangerous shaft in one corner. In areas where the contours were suitable, mines could be drained by tunnels or soughs driven into the workings from a lower contour in an adjacent valley. In Derbyshire, the first sough for which documentary evidence is available was commenced in 1632 by the Dutch engineer, Sir Cornelius Vermuyden, who was to be employed by the Earl of Bedford to direct the drainage of the fens. He constructed a sough to drain the Gang Vein between Cromford and Wirksworth, a process carried further by the driving of the Cromford Sough in the last years of the seventeenth century. At least 150 soughs are known for certain, although twice that number are referred to in documentary sources.[2] With larger mining companies involved, soughs could be driven to drain several mines rather than individual ones, such as the Hill Carr Sough, begun in 1766 and driven 700 feet below the surface of Stanton Moor to drain the Alport Mines. Four miles in length, it took twenty-one years to complete and paid for itself in increased output within two years of completion. At the surface, the entrances to many of these soughs still exist along river valleys in Derbyshire and the water gushing out of them plays an important part in modern domestic supply.

Soughs were constructed elsewhere in the country, although the greatest concentration of them is to be found in Derbyshire. Mine captains often introduced their local practices in other areas, such as Cornelius Flint, mineral agent to the Duke of Devonshire, who constructed the Duke's Level to drain the mines of Grassington Moor on his estates in Yorkshire. In Flintshire (now Gwynedd) the Grosvenor family started a level in 1818 to drain their lead-mines on Halkyn Mountain and this Deep Level was not completed until 1901. In Cornwall, too, mining was assisted by the construction of drainage levels, notably the Great County Adit, started in 1748. It eventually totalled around 30 miles and drained forty-six mines. This particular drainage system was largely responsible for the great increase in copper output from Cornish mines in the second half of the eighteenth century.

Not all mines, however, were situated in terrain suitable for the construction of drainage levels, and anyway these could only drain mines to the depth of the adjacent river valleys. The use of water-power to drive pumps was, like drainage levels, limited by the terrain in which the mines were situated and its use could be hindered by both drought and frost. Steam engines were independent both of the topography, as can be seen by the Cornish examples situated on cliff edges, and of the weather, but needed to be kept supplied with coal. Consequently their use was more common in mining districts situated close to coalfields, as in the Pennines and North Wales, or in the case of Cornwall, where coal was brought by sea from South Wales as a return cargo for copper ores being taken there for smelting. In Cardiganshire, on the other hand, water-power continued to be used until, in many cases, the closure of the mines. Steam engines were erected, but fuel economy was obviously more important on metalliferous than coalmines and many were abandoned as being too costly to run.

Whether water- or steam power was used for draining the mines, the effect on the landscape was more dramatic than had ever been caused by hand pumps or drainage soughs. To obtain sufficient water for both power and dressing purposes, streams had to be harnessed over a wide area and vast networks of leats constructed to bring water to storage ponds. In the neighbourhood of Allenheads in Northumberland, worked by the Blackett-Beaumont Company, ten reservoirs were built with a total capacity of over 60 million gallons. The water from these both powered a large number of surface and underground pumping engines and was used on the dressing floors. The outflow from the Corbitmere Dam, built in the Rookhope Valley, fed the Allenheads dressing floors through a series of shafts and levels, then powered four underground water-wheels and, above ground again, four corn mills before providing water for the Allen Smelt Mill north of Allendale Town.[3] Contouring could be avoided if aqueducts were used to carry water across valleys, although the cost of these prevented their widespread adoption. The most spectacular example to survive is probably the Luxulyan Viaduct in Cornwall, constructed between 1839 and 1842 by Joseph Treffry. This carried both a tramway and an aqueduct 100 feet above the valley, the latter conveying water to a china stone crushing mill.

Few of the water-wheels themselves survive and, inevitably, the best known is the most spectacular of them which is not typical of the majority. The Lady Isabella wheel, named after the wife of the island governor, was designed in 1854 by Robert Casement to drain the Laxey lead- and zinc-mines in the Isle of Man. A pitch-back wheel 72 feet in diameter and 6 feet wide, it drained a 1,200-feet deep mine shaft by means of flat rods. It was always a tourist attraction and when the Laxey Mines closed in 1929, the wheel and flat rod system were preserved first in private hands, and then by the Isle of Man government (Plate 14). More typical remains of water-powered mine drainage systems can be found in Cardiganshire, where wheelpits or wheelcases (so called if they stand above ground) for wheels 30 to 40 feet in diameter are still visible on many mine sites. At Esgair Hir, a 40-feet diameter pumping wheel stood in the valley bottom and pumped a shaft on the hill top via a line of flat rods two-thirds of a mile long, which entered the shaft through a tunnel: the rods were kept in tension by balance boxes both at the shaft and wheel ends. The flat rods no longer survive, but along their route obstructions had been trenched through and remains of the wooden stands supporting the rods can still be found.[4] Lines of these stands can be traced on other mine sites, such as Graiggoch in the Cwmnewydion Valley, where a row of wooden supports can be seen over a distance of 400 feet between the wheelpit and the shaft: stone supports can be seen outside Bolton Gill Level near Grassington.

The Newcomen engine, widely used for draining coalmines from the first years of the eighteenth century, was not so popular on metal-mines because of its large coal consumption. Nevertheless, if coal was available, the Newcomen engine could help free a mining company from the unreliability of water-power. In Flintshire, a Newcomen engine was erected at the Talargoch lead-mines at Dyserth in about 1715, only three years after the first engine had been erected at Dudley Colliery. Few were introduced into Derbyshire and the excessive consumption of fuel by the atmospheric engine was always a great drawback in Cornwall. Five are recorded as having been introduced by 1727, including one at Wheal Rose where the adventurers found the costs so prohibitive that they preferred to drive a 1.5 mile adit so that they could dispense with the engine. The import duty levied on all seaborne coal held back Cornish mining, but in 1739 coal landed in Cornwall for the purposes of operating engines was exempted. There followed a spate of engine building, so that by 1778 over sixty Newcomen engines had been erected.[5] Carriage costs of coal were still high, but Cornish mine captains working copper had to accept these in face of competition from the deposits of copper ore discovered in 1768 at Parys Mountain in Anglesey. But when

Plate 14 The Lady Isabella water-wheel in the Isle of Man. This was built in 1854 to pump the Great Laxey lead-mines. Water was collected from the Laxey Valley, stored in a cistern and fed by underground pipeline to the top of the huge wheel. A crank arm on the wheel was connected by a long flat rod system supported by rollers on the bridge seen in the background to pump rods in the Engine Shaft. Such was national pride that a casting of the Manx emblem was placed on the end of the wheelcase.

improved engines with a reduced rate of fuel consumption became available, the Newcomen engines were rapidly abandoned.

The economy of fuel consumption of Watt engines was invaluable to metal-mining companies who relied on steam power. On North Consols Mine at Gwennap in Cornwall, seven Newcomen engines had consumed over 19,000 tons of coal in 1778–9, whereas the five Watt engines which had replaced them by 1783 consumed less than one-third of this. By that year, no working Newcomen engine survived in Cornwall, the result of the installation of twenty-two Watt engines in the six years since the first one was started at Chacewater Mine in 1777.[6] Cornish mine captains came to prefer an improved Watt engine, consisting of one large, often 80-inch diameter, single cylinder using high pressure steam acting expansively on one side of the piston only. This, combined with the use of plunger pumps in the shaft, gave them the power to pump from depths of over 2,000 feet. They are often known as 'Cornish' engines because of their widespread use in the county. The tall engine houses had to be strongly constructed both to support the great pumping beam and to hold down the cylinder against the reciprocating motion of the piston and pump rods. A balance bob, or weighted box on a pivot, helped counteract the tremendous weight of the pump rods, and the pit for this can still often be found close to the pumping shaft of a mine. In Cornwall, granite was preferred both for the foundations and the walls, quarries often being specially opened for the construction of a particular engine house. The bob-wall was usually 5 to 6 feet thick, the other walls about 3 feet. Inside, massive blocks of stone supported the cylinder, with a square hole or well in front of it to contain ancillary equipment such as the condenser and air pump. The position of bolt holes in the cylinder blocks of surviving engine houses indicate the size of the cylinder last mounted there.

The boiler house usually adjoined the engine house with a chimney stack built into the back corner to assist stability. The stack was usually circular in section, built of stone with upper courses of brick. Most engine houses were built to a fairly standard pattern and had few architectural features. The windows often had brick lintels, but were kept to a minimum size to avoid weakening the structure. Some houses had numerous slit windows instead of a few larger ones, such as that for an 80-inch engine at Tregurtha Downs, near Marazion. This engine, originally made at the Copperhouse Foundry in Hayle in 1854, worked on two other mines before being moved to Tregurtha Downs in 1882. A photograph survives of the arrival of the bob, or beam, weighing 40 tons, which was dragged to the site by forty-five horses, watched by over 1,000 people. Such scenes must have been quite common in nineteenth-century Cornwall, since engines were frequently purchased second-hand from bankrupt mines and installed elsewhere. The Tregurtha Downs engine, in fact, went on to South Crofty near Camborne in 1903, where it remains in the care of the National Trust.

The Watt pumping engine was never as dominant in other metal-mining areas as it was in Cornwall. In Derbyshire, a lot of reliance was placed on the sough system, combined with underground hydraulic engines, and it was often engineers from outside the county who advised the use of steam engines. For example, the Cornish engineer John Taylor reopened the Magpie Mine near Sheldon in 1839 and replaced the earlier Newcomen engine with a 40-inch Cornish engine. The cylinder size was increased to 70 inches in 1868, but not even this engine could pump the mine dry from a depth of over 700 feet, and the Magpie Sough was constructed between 1873 and 1881.[7] The surviving engine houses at Magpie Mine, in the care of the Peak District Mines Historical Society, are therefore completely atypical of the Derbyshire mining scene. By contrast, the Shropshire lead-mining area, close to a coalfield, made far more extensive use of steam; engine houses were as thickly concentrated here as in Cornwall, although confined to a very much smaller area. In Flintshire, the Cornish engine at City Shaft, Minera, pumped water from a depth of 1,220 feet, the deepest of all the North Wales metal-mines, and the engine house has been restored as part of an interpretation scheme for this rich lead-mining area. Isolated pumping engine houses survive in other mining districts, notably the Isle of Man and Cardiganshire, but they were frequently used in conjunction with water-power and so were never built in the concentrations of Cornwall or Shropshire.

The raising of men and minerals from the mine by mechanical means exercised the minds of mine captains rather less than the perennial problem of keeping the workings free of water. Simple windlasses or horse-gins were used to raise ore, while the miners themselves usually had to descend and ascend by a series of ladders leading to wooden platforms set in the side of the shaft. Occasionally, separate climbing shafts were provided, of narrow diameter with wooden poles or stemples fixed diagonally across to form footholds. These tiny shafts, often unmarked, still exist in mining areas, especially in Wales and parts of west Cornwall. The climb 'to grass', as the surface was called, often by 1,500 feet of ladders, was a cruel necessity after a shift spent in bad air and high temperatures. Following their introduction in Germany, man-engines were installed in Cornwall. These had one or two reciprocating rods in the shaft, similar to pump rods, with small platforms which men stepped on and off to descend and ascend the mine shaft. They were far from accident free: the platforms became covered in mud and grease and the actual process of stepping on and off became mesmeric, but accidents were probably fewer than those caused by exhaustion in the ladder shafts. At Levant Mine, on the extreme west coast of Cornwall, an adit entered the workings from the cliff face, set above the great chasm known as Levant Zawn, and miners had to climb the cliff as well. This mine was worked to a depth of over 1,800 feet and it is hardly surprising that a man-engine was eventually installed which made use of two sets of rods. Sixteen of these man-engines were commissioned in

Cornwall, the first in 1842, all in large mines working to considerable depths. All the man-engines had gone out of use by the end of the century and only Levant remained. In October 1919, the rods broke away suddenly at a time when a full shift of about 150 men were ascending on it to the surface.[8] It was the only instance in Cornwall of the failure of man-engine rods involving fatalities, and it is sad that an attempt to ease the lot of miners ascending to grass should end in this way.

The rotative steam engine was first utilised for winding in Cornwall only two years after its invention in 1782. The steam 'whim' was never as common in other mining districts, where horse- or water-power continued to be used for raising ore, but in Cornwall's deep mines it was a necessity. An early example survives at Levant in the care of the Trevithick Society and the National Trust. A small, squat engine house contains a 24-inch whim engine built by Harveys of Hayle in 1840, which worked continuously raising ore for ninety years until the mine was abandoned in 1930, but has now been restored to working order (Plate 15). A later example, also in the care of the National Trust, stands beside the A30 in Pool and was the last rotative beam winding engine built in Cornwall, being made at the Holman Foundry in Camborne in 1887. Both these surviving engines help to demonstrate the difference between the houses for pumping and winding engines. Generally, the cylinders for the latter were much smaller, usually about 22 inches in diameter by the end of the nineteenth century compared with 80 or 90 inches for a pumping engine, and so whim houses were smaller and less robust. The beam operated a sweep rod connected to the winding drum shaft by a crank, and one or two large flywheels were needed to ensure continuous motion. Generally, the winding drum and flywheel were housed on substantial stone foundations or 'loadings' in front of the engine house. These often survive long after the machinery has been scrapped and it is possible to work out the layout from the slots cut in the stone for the flywheel and crank, and from the bolts which held the winding drum. To avoid cluttering the top of a shaft, whims were usually placed some distance away and the ropes were sometimes carried to the shaft on pulley wheels mounted on a series of tall supports.[9] The stark beauty of abandoned mining setts represents only the skeleton of an ephemeral nineteenth-century landscape of wood and galvanised iron, which has disintegrated leaving only the stone or brick structures behind.

Ore was originally raised in baskets or wooden and leather buckets, but by the nineteenth century these had been displaced by the kibble, a wrought iron bucket which was egg-shaped to avoid damage to the shafts. From the middle of the century skips travelling on guides in the shaft were introduced. These were used for winding men as well as ore, as is shown in the famous painting by C.J. Hook of the diagonal skip road at Botallack Mine on the Cornish coast, exhibited at the Royal Academy in 1864. A year later the Prince and Princess of Wales descended this skip road, the

Plate 15 A *c.* 1900 photograph from the west of the Levant Mine near St Just in Cornwall. In the left foreground is the 1840 winding engine house, complete with its boiler house and chimney, in front of the twin-wheeled headgear over Skip shaft. In the centre is the pumping engine house, also with boiler house and chimney stack, which worked to the Engine Shaft, very close to the winding shaft. The third engine house in the right background drove the stamps which crushed ore brought by incline up from the winding shaft. The fourth chimney stack, on the cliff edge, belonged to an arsenic calciner located on the down slope beside the dressing floors to the east of the stamps. Other buildings on the right are the count house and the miners' dry, where they changed and dried their clothing. The winding engine has been restored and is regularly steamed.

timber staging for which ran down the face of the cliff, and this part of Botallack Mine was known thereafter as the Crowns Section. Closed in 1895, the surviving engine houses are a spectacular feature of the Cornish coastal landscape. Where access to mines was by adit rather than shaft, the ore was 'trammed' out in small trucks on wooden, later iron, rails. These were usually pushed manually rather than horse-drawn, since metal-mines were rarely large enough to permit the use of horses underground.

Dressing the ores

The ore was stored in bowl-shaped stone hoppers or bouse teams which are a feature of many lead-mining sites. ('Bouse' is a term meaning mixed ore and stone.) Since the miners worked as independent teams and were paid by output, the ore raised by the different teams had to be stored separately, hence the long lines of storage hoppers at Killhope in Weardale and alongside Gunnerside Gill in Swaledale. By the mid-nineteenth century, some of these hoppers were made larger and used for the initial washing of the bouse, and the leats bringing water to the rear of them can often be traced. These 'wash kilns' are a feature of many Cardiganshire mines, notably Bryn Dyfi, Bron Floyd and Ystrad Einion.[10] The ore then had to be broken up to free the wanted minerals from the surrounding 'gangue' or veinstuff. How this was done depended very much on the mineral: cassiterite, the chief ore of tin, is disseminated finely throughout the gangue and so has to be crushed very thoroughly to free it; galena, on the other hand, the chief ore of lead, can often be picked out in lumps and the gangue does not need to be reduced to so small a size as tin ore to free it. The initial sorting and breaking, known as 'ragging', 'spalling' or 'bucking', was generally carried out by boys or women on a flat floor with a heavy flagged surface. The Cornish 'bal maidens', as they were known, were a formidable body of strong-armed women. Their manual labour was later lessened by the use of stone breakers or jaw crushers, which were in use in most large mines by the close of the nineteenth century.

The finer crushing needed more than manual effort, and in the eighteenth and even into the nineteenth century in some smaller mines, horse-power was utilised. A circular stone turned on a horizontal axle fixed to a central post which rotated on a bearing. The axle was drawn round by a horse, the veinstuff being laid in the path of the rolling stone. These circles can often be detected in the modern landscape, most easily, perhaps, at Odin Mines in Derbyshire where a circular iron track and iron-tyred stone roller survive *in situ*. Circular tracks made up of flat stones remain at Eldon Hill, Watts Grove and Little Pasture Mines, also in Derbyshire, sometimes associated with fragments of the roller stone. Water-power had also been used to drive crushing machinery from at least the sixteenth century. Stamp mills were described and illustrated by Georgius Agricola in *De Re Metallica* of 1556 and were already in use in England. They consisted of a number of wooden rods or pestles, later shod with iron, which were alternately raised and dropped on to stone blocks or mortars to pulverise the ore. Many of these blocks, often in association with wheelpits and leats, have been found on Dartmoor. With some modification, particularly the use of iron instead of wood, water-powered stamps remained in use in some areas into the twentieth century. At Bonsor Mine in the Lake District in the 1840s, it was said that the rhythmic sound of the stamps could be heard in the village night and day except on Sundays.

Stamps were not a common feature of lead- or copper-mines except where the ore was contained in a hard matrix such as quartz, but they were in general use for tin in south-west England. Larger mines adopted steam to power their stamps from the beginning of the nineteenth century: by the end of the century a single engine could drive as many as 140 head of stamps. The batteries of stamps were usually arranged symmetrically on either side of a rotative beam engine house and the massive bearing supports for the stamps' axles may also remain. The ore was crushed in water, and some stamps engines also worked pumps either to raise the necessary water from adits or to return it from a cistern at the bottom of the dressing floors. Engine houses which contained this type of engine can often be detected by the presence of a second thick bob-wall at the rear of the house. Good examples of stamps engine houses survive at Wheal Peevor near Redruth and at Wheal Grenville on the Great Flat Lode. More efficient stamps were developed after the 1850s. These worked faster than the traditional stamps, but their use was not widespread because the capital costs required for installation could not be justified at a time when the whole profitability of Cornish mining was in question.

Lead and copper ores, needing less thorough reduction, were better crushed between rollers, and from the early nineteenth century water-powered crushing mills could be found at many mines. None survive intact, but at Killhope in Weardale the massive 33-feet diameter wheel with an extensive launder bringing water to it, gives some idea of their appearance. Elsewhere the roofless crusher buildings stand alongside their empty wheelpits, massively built to resist the vibrations of the cast iron rollers. The sockets for the transverse beams which supported the rollers may be found in the front and back walls.[11] The rollers were held against each other by heavy weights suspended on the end of long arms which often projected outside the building: massive rocks with iron lifting eyes can occasionally indicate the site of a former crusher house. Since crushing mills often stood close to other wheels used for pumping and winding, the water-power installations must have presented a spectacular sight.

After crushing or stamping, the vein material passed through a bewildering variety of dressing processes designed to separate the ore from the gangue minerals. The principle common to all of them is that the metal ore is more dense than the unwanted minerals and so, if the crushed material is floated or agitated in a trough of water, the ore separates out first and falls to the bottom. This is most easily achieved by the use of a gravity flow system on a sloping site with the stamps or crusher at the top and the dressing apparatus arranged in a sequence below them. Since much of the equipment was made of wood, little usually survives, but a common feature was the circular round buddle base, either convex or concave, built of stone or concrete and sometimes wooden-lined: this was used for the separation of the finer particles of ore.

By the end of the nineteenth century, more elaborate mechanically-powered separators were used as hand labour became more costly, and large concentrating mills, often roofed with corrugated iron, became a feature of mining districts. An old view of the tin-dressing mill at Cooks Kitchen Mine near Camborne in Cornwall illustrates the unsightly nature of the often ephemeral buildings containing the dressing plant, some parts of which could well be disused and decaying because a new method was being tried out in a hastily constructed building (Plate 16). Of particular importance was the introduction of the flotation method around 1900 for dealing with complex ores. The powdered ore was treated with specific chemicals which allowed differential flotation in water, enabling the various minerals to be separated. The many remains of dressing floors which survive in today's landscape are but ground-level indications of a vast superstructure which grew up unplanned and disappeared equally quickly when the mine was disused, leaving only foundations or structures set in the ground. It is, however, possible to recreate earlier mining landscapes from surviving field evidence and to distinguish both different periods and the treatment of different ores.

Lead-mining landscapes

Looking first at lead, the most widespread of our past metal-mining industries, the appearance of the eighteenth-century lead-mining area is perhaps best represented in Derbyshire and the northern Pennines. The mines were small and shallow, wound by human or animal power and drained by means of soughs or water-wheels, or perhaps by Newcomen engines in mines owned by wealthy partnerships. Dressing was labour-intensive, the ore being stored in bins and then broken up with hammers on bucking floors. Further crushing would be either by means of a horse-powered edge crusher or by use of water-powered stamps. Simple hand dressing was carried out in rectangular buddles and wooden kieves. No site survives intact from this period, but smaller mines continued to use the old methods in the nineteenth century. The hillocks of bell pits are still a common sight in Derbyshire, and horse whim circles and bucking floors can be detected near them by the discerning eye. The packhorse routes used to take concentrated ore to smelting mills can be determined both from maps and in the field.

The more intensive, highly capitalised lead-mining of the nineteenth century has left clearer traces in the modern landscape. In water-powered areas, the most spectacular features are usually the massive wheelpits or wheelcases which housed overshot wheels for pumping, drawing and crushing, like the 40-foot wheelcase at Temple Mine. This dramatic site near Ponterwyd in Dyfed also retains its dressing floor. A well laid-out site is that of Bryn Dyfi, also in Dyfed, now scheduled as an Ancient Monument. It was designed by D.C. Davies and his son, authors of several books on mining technology, and came into operation in 1881. Unfortunately, the surface installations had preceded intensive underground exploration, and the mine closed within two years after producing less than 30 tons of lead ore. Consequently, because it was little used, the surface equipment has survived intact. The two shafts and adit of the mine itself

Plate 16 The ephemeral landscape of Cornish mine ore dressing can be appreciated in this *c.* 1890 photograph of Cooks Kitchen Mine in the Tuckingmill Valley near Camborne. The sloping site down to the Red River was utilised to give a gravity flow between the machines. There is an extensive range of round buddles in the foreground with horizontal frames for treating slimes on the right. On the horizon are several engine houses, headgear, miners' dry and count house. By 1865 there were four steam engines and six water-wheels in use on the sett and a man-engine was later added.

are about 1,000 yards distant from the dressing floors and connected to them by a gently contouring tramway along which trucks were probably pushed by hand before being tipped into two wash kilns. Three reservoirs were constructed well above the site, fed by leats from the hillside and bringing water to the dressing floors by leat and launder. Two wheelpits and much of the dressing plant remain. The whole plant was well constructed of solid masonry, and typifies a small but well-planned layout.[12] A far larger dressing floor is that at Killhope, first constructed to process ore from the Park Level Mine in 1853 and modernised twenty years later. This has been excavated and the full lead dressing process recreated for the benefit of visitors to this remote Weardale site.

In lead-mining areas close to coalfields, the nineteenth-century landscape would have presented a very different appearance, being dominated by tall engine houses rather than graceful water-wheels. However, the very accessibility of these areas, which enabled coal to be transported there, has also led to wholesale demolition of the machinery for its scrap value or to vandalism, with the result that no steam-powered lead-mining site survives as intact as its water-powered equivalent. At Snailbeach in Shropshire, held to be the richest mine per acre of ground in Europe, the chimney of the Lordshill New Engine Shaft still dominates the scene, but the productivity of the mine is expressed more in the vast spoil heaps than in surviving buildings, which include pumping, winding and compressor houses and a railway system. Many other mines in the vicinity still retain their brick- or stone-built engine houses, but little other surface evidence. Many worked-out lead-mines were given a new lease of life in the last decades of the nineteenth century through the demand for zinc sulphide or blende. Zinc had earlier been obtained from calamine, zinc carbonate, which was mined in the Mendips and the Bristol region, hence the establishment of the important brass industry in Bristol. Many former lead-mines had their waste tips reprocessed for blende, since the introduction of the flotation process was particularly suitable for zinc ore separation. New dressing floors were constructed especially for this purpose and their concrete remains are a feature of many Welsh mining sites, particularly Nantymwyn near Lampeter, Gwaithgoch in Dyfed and Hafna in the Gwydyr Forest. In the recent past, fears of contamination from heavy metal waste have resulted in extensive clearance and reshaping of such areas and at Minera in Clwyd the newly-consolidated foundations of buildings now sit oddly in a grassy landscape.

Copper- and tin-mining landscapes

The methods of winning and dressing copper ores closely resemble those of lead. The remains of the water-powered phase of the industry can be found in North Wales, particularly in the old county of Caernarvonshire where mines were located along river courses high up in the mountains. In Cwm Pennant, to the north of Portmadoc, the Cwm Cipwrth Mine preserves its water-powered winding and pumping gear, the wheel for which came all the way from Truro to be erected on a remote site at 500 feet above sea level. Cornwall is usually thought of primarily as a tin-mining area, but in fact reached the height of its industrial development during the copper boom of the middle years of the nineteenth century, when over 600 steam engines were at work in the county and scores of foundries came into existence to manufacture the giant engines needed to pump out the deep mines. The reconstruction of the dressing floors for tin ores in the late nineteenth century has obliterated most of the evidence for the copper-mining period.

Copper output in Great Britain had been revolutionised in the mid-eighteenth century by the discovery of vast deposits of copper in Parys Mountain on Anglesey. In the 1780s its production was greater than that of any other mine in Europe, amounting to over 3,000 tons a year. The ore was first worked by means of shafts and levels, but eventually the whole mountain became a vast pit, 70 feet deep, worked by opencast methods. The pit was pumped out first by a five-sailed windmill, whose remains can still be seen, and later by a steam engine. The water was rich in copper sulphate, and the copper was recovered by pumping the water into a series of pits. Scrap iron was deposited in these, to be dissolved by the copper sulphate, leaving a copper-bearing sludge on the bottom of the pits. Similar precipitation pits can be found on other copper-dressing sites in Britain.

Tin has been extracted in south-west England since prehistoric times, not at first by mining but from surface deposits where the parent rock had been broken up and tin-bearing gravels carried away by water. From the medieval period, tin streamers dammed watercourses and constructed holding ponds to settle out the ore. This has resulted in areas of broken ground, which can be seen very well at Beckamoor Combe on the southern slopes of Dartmoor and at Butterne Hill on the northern slopes of Bodmin Moor. The raw material for the tin streamer was increased once the waste from dressing plant attached to inland mines was discharged into the rivers. Elaborate equipment was set up along streams like the Red River between Camborne and Redruth, designed to extract the last ounce from the waste slimes.

The intense exploitation of Cornish minerals during the nineteenth century has left few traces of previous workings. The tin-mining landscape is mainly a nineteenth-century one, its most characteristic feature being the engine houses which were built for pumping and winding the mines and crushing the ore. Pairs of engine houses, pumping and winding from the same mineshaft, are a common feature and good examples survive at Pascoe's and Lyle's shafts on the Basset Mines, at Wheal Uny south of Redruth and most spectacularly at Trewavas Head, Botallack and Wheal Coates on the coast. More rare are the trios of engine houses which include a stamps engine, as at Wheal Peevor, separated from Redruth by the new bypass. Several dressing floors remain because of their concrete construction, including the particularly well-preserved example at West Basset, near the village of Carnkie, below Carn Brea. The plant was built on a slope, the stamps engine powering ninety-six heads of stamps and also pumping water back from the dressing floors. Below this is a series of buildings where ore dressing was carried out, representing the last in a whole series of modifications occurring between 1876 and 1906 in response to new developments in the tin-dressing process.[13]

A building unique to Cornwall is the calciner, designed to heat the partly dressed ore to drive off unwanted impurities such as sulphur and arsenic. The ore was heated on a slowly rotating iron plate, and the circular interior of the small, squat, square building indicates its function. The vapour passed through a flue to a chimney, but on some mine sites the arsenic was recovered as a saleable commodity. This was done by passing the vapour through a labyrinth of flues on the inside of which the arsenic condensed and could then be collected. A particularly good example of this kind of calciner survives at Botallack Mine on the west coast of Cornwall, but these labyrinthine flues form a major landscape feature on many sites in the Camborne–Redruth area.

The mining of gold has left few traces on the landscape of Britain. The mine at Dolaucothi in Dyfed, now in the care of the National Trust, has probably the longest history of any mine in Britain, with traces of Roman working, including aqueducts for hushing, and finally the establishment of a twentieth-century dressing plant. More important during the nineteenth century were the hills just to the north of the Afon Mawddach in Gwynedd. The discovery of ore in the 1840s prompted a minor 'gold rush' to the area, and many trial levels and shafts were dug. One of the most productive mines was Gwynfynydd, where the dressing mill was powered first by a water-wheel and then by a turbine, the rock-cut leat for which can still be seen.

Smelting the ores

In order to obtain the pure metal, the concentrated metallic ores had to be smelted. This had as dramatic an effect upon the landscape as mining, partly in the construction of furnaces and other buildings, but even more in the devastation of vegetation caused by the noxious gases given off in the process. There are significant differences in the smelting methods for the individual non-ferrous metals, but in general they followed a similar evolution because of changes in power and fuel sources. The earliest furnaces were the bole hills, which relied on natural draught to raise the temperature of the fuel and ore to the melting-point of the metal. These were gradually replaced in the early modern period by ore hearths, using water-powered bellows to induce the necessary draught. In both these types of smelter, the fuel and ore were intermixed and so coal could not be used as a fuel as it would contaminate the metal. As a result, charcoal, produced from either wood or peat, and white coal, a kiln-dried wood, were the main fuels used. By the eighteenth century, the shortage of wood encouraged the use of coal and a new kind of furnace, the reverberatory, was developed in which the burning fuel and ore were kept separate. A chimney was used for inducing the draught rather than bellows.

Bole hills were situated on hillsides and usually aligned with the prevailing wind. Their former presence can often be detected by slag scatters and burnt stones, such as those on the slopes of Calver Hill above Heelaugh in Swaledale, while the name 'bole hill' on maps also indicates a former smelting location. When water-powered bellows were introduced, the furnaces had to migrate to the river valleys, but the availability of wood and peat for fuel was also important and so ore hearths could not be built in every metal-mining area. The introduction of the coal-fired reverberatory furnace meant that the costs of transporting ore to fuel or vice versa had to be balanced, and although these furnaces were constructed in the mining areas they were more often sited away from the mines near suitable coalfields.

The tin-mining industry of Devon and Cornwall had been governed by Stannary Laws since the thirteenth century and all smelting took place within the two counties. This was because of the coinage system, by which the tin ingot had to be 'coined' or sampled for assay and stamped before sale. In the early modern period, the ore hearth was contained in a blowing house usually associated with the tin streaming sites. The ore produced from streaming was exceptionally pure, and being an oxide of tin was easily reduced to grain tin without producing any noxious gases. The small stone-built blowing house with a thatched roof frequently also accommodated, in addition to the water-powered bellows, primitive stamping and dressing machines. Several examples of blowing houses may be found on Dartmoor, for example, at Merrivale where the remains of three survive, all powered by the River Walkham. Apart from the wheelpits, prominent features of these structures are the granite ingot moulds into which the molten tin was run and the hollowed out mortar stones from the stamps. Blowing houses gradually went out of use once lode-mining began, but one survived in use at St Austell until the 1860s.

The more complex lode tin ore was treated in reverberatory furnaces and large smelting plants were constructed near the coinage towns, particularly Truro and Penzance where coal from South Wales could easily reach them. Ore was sold from the mines at monthly ticketings to the highest bidder for smelting, and the financial dominance achieved by the smelters eventually enabled them to control the mines themselves. Very little remains of these reverberatory furnaces apart from some buildings and chimneys at Seleggan, near Redruth, which opened in 1887 and was the last to close in 1931. Coinage ceased in 1838 and although local smelting continued, the lack of coal in the south-west, combined with growing tin ore imports, encouraged the Cornish smelting companies to build their furnaces elsewhere. An extensive tinplate industry grew up in South Wales, where iron sheets were coated with molten tin in foundry-like buildings with tall chimneys above each of the tin baths. Tinplate works were often added to strip rolling mills, but there are few remains apart from the mill now preserved as a museum to the tinplate industry at Kidwelly.

Copper ores were considerably more complex than tin, being mixtures of the sulphides of both copper and iron which required repeated smelting to refine the metal. The melting-point of copper was also much higher than tin, and so the smelting process was expensive in fuel. Only after the repeal of customs dues on ship-borne coal landings in 1710 were any copper smelters established in south-west England and these were confined to the west of Truro where most of the copper ores were produced. The last to operate was the Cornish Copper Co. at Hayle; this opened in 1756 but was forced to close by 1819 due to increasing transport costs and the competition from the Parys Mountain copper deposits on Anglesey. There is little evidence of its existence apart from the slag blocks used to construct the walls of the Copperhouse Dock in Hayle. It proved more economic to take the ore to the coalfields as return cargo for the increasing coal imports required to fuel the growing number of steam engines in the mines of the south-west.

Reverberatory furnaces were constructed at Bristol, in Lancashire and especially in South Wales, where by the 1850s the Swansea Valley was the world centre for copper smelting. An engraving made by a French geographer in 1865 gives some idea of the intensity of the development, with integrated smelting and rolling mills filling the Tawe Valley. Copper smelting here ceased in the 1920s and because of the reclamation of the devastated landscape in the Swansea Valley Project, few buildings remain. The Hafod Works, established in 1810 by John Vivian, retains two engine houses to drive rolling mills and a locomotive shed. Several examples of the use of copper slag blocks may be found on the site and in the company houses at Vivianstown on the nearby Neath Road. The adjacent Morfa works, also established by a Cornishman, possess some buildings from the smelting works as well as the river quay where the ore was unloaded.[14] Copper ores from Anglesey were smelted at St Helens in Lancashire and the ingots processed at water-powered rolling mills in the Greenfield Valley north of Holywell in Clwyd, where their remains have been consolidated.

Unlike copper, lead is a relatively simple metal to smelt, having a low melting-point which enabled low grade fuels such as peat to be utilised. The galena ore was a sulphide and the smelting process generated large quantities of sulphur dioxide fumes and some vapourised lead. This poisoned vegetation in the vicinity of the smelt mills which were consequently built well away from settlements. The ore was smelted on a hearth similar to a blacksmith's hearth, which was placed under a stone arch connected to a flue. Liquid lead ran off into a heated vessel or sumpter pot, from which it was ladled for casting. The ores were often roasted before treatment, and since the slags from the smelting process could contain up to 20 per cent lead and were worth re-treating, there were often several hearths in the smelt mill. The restored smelt mill at Wanlockhead in Scotland has examples of both types of hearth. The air blasts were provided by water-powered bellows, and so a typical smelt mill can be recognised by its wheelpit, bellows room, hearths under arches and flue system. Huge open-sided covered barns for drying and storing peat were especially built nearby. The best surviving examples are be found in the Yorkshire Dales National Park: the Grinton, Marrick, Old Gang and

Surrender Mills in Swaledale.[15] All of these exhibit the lengthy condensing flues added in the nineteenth century, which led to chimneys on nearby hills and were designed both to condense vapourised lead and to disperse noxious fumes. The flues were often in pairs, so that one set could be closed off with stone slabs for cleaning, and sometimes incorporated condenser chambers through which water flowed to aid precipitation. Splendid examples of these can be seen on Yarnbury Moor above Grassington in Wharfedale and near Allenheads in Northumberland.

The reverberatory furnace, which made use of coal as a fuel rather than peat, was never popular in the Yorkshire Dales, but was introduced into Derbyshire in the 1730s and other lead-mining areas near coalfields, such as Shropshire. There are fewer remains of these than of ore hearths, but the restored chimney on the Stonedge cupola in Derbyshire, dating from *c.* 1770, is probably the oldest surviving industrial chimney in Britain. The labyrinthine complex of flues at Charterhouse-on-Mendip in Somerset was used for re-smelting ancient lead slags. An equally spectacular survival of the lead industry is the shot tower at Chester, built in 1799–1800, one of many which once existed to manufacture lead shot by dropping molten lead the full height of the tower into water.

Metallic zinc was first produced commercially in Bristol by William Champion in the 1740s, but large-scale production did not begin until the development of the galvanising process in the 1840s enabled large quantities of corrugated iron sheeting to be produced, much of it sent out to the colonies. The metal had found an earlier use in the manufacture of brass, an alloy of zinc and copper. This too was centred on Bristol, where brass utensils were formed under water-powered hammers in a 'battery' works. These were heat-treated or annealed in tall tapering furnaces which had arches front and rear; the hearths were loaded through the front and fired from the rear. A long vertical slot in the front wall accommodated a pivoted wooden beam upon which the loading door was hung. The outer walls of two of Champion's annealing furnaces of the 1760s can still be seen at Kelston on the River Avon, together with some workers' housing, but the most complete works is that at nearby Saltford. This was a former fulling mill which turned to brass-making in 1721 and continued to operate into the 1920s.[16]

In general, non-ferrous metal-mining and processing were carried out in areas since untouched by twentieth-century development and so much can be learned from the field evidence. Dramatic changes in topography were caused by the diversion of streams and the creation of artificial watercourses and ponds as well as by opencast mining and the deposition of tailings as waste dumps. Features were also added to the landscape in the form of wheelpits, engine houses, dressing floors, furnaces, roads, tramways and ancillary buildings. Ferrous metal-mining, on the other hand, although more widespread, has left fewer identifiable features, largely because iron extraction continued until at least the middle of the twentieth century and so obliterated earlier developments. Fewer buildings survive than in non-ferrous mining areas, but the landscape changes have been, in places, equally drastic.

Iron-mining and smelting

Nodular iron ores were often mined by means of bell pits similar to those of the lead-mining industry but not positioned in regular lines since iron, unlike lead, does not occur in veins. At Bentley Grange, 6 miles north-west of Barnsley, are several fields pock-marked with the spoil heaps of iron-mines. By the nineteenth century, when it became possible to utilise the lower grade Jurassic ironstone, large opencast quarries were developed from which several feet of ironstone were extracted before the land was returned to agriculture. Consequently, many roads and isolated farmsteads in the East Midlands now stand above the level of the surrounding fields and are an interesting example of the way in which past industrial activity can transform a landscape yet leave no trace of itself behind. The nineteenth-century iron industry was based upon an extensive narrow and standard gauge railway network and abandoned tracks, inclines and bridges enable the former lines to be traced. A few survive: the narrow gauge Ravenglass and Eskdale Railway, now a tourist attraction, was constructed in 1875 to bring iron ore from the Nab Gill Mines in Eskdale to the Furness Railway at Ravenglass, and in both east Leicestershire and Northamptonshire enthusiasts still operate steam locomotives on sections of track built to serve ironstone quarrying.

The haematite ores of the Lake District were mined rather than quarried, the result being considerable surface subsidence leading to disturbed ground and the formation of pools of water. The numerous engine houses, shafts and engine mountings associated with the mines have generally been obliterated, even at the productive Hodbarrow Mine which did not close until 1968. What has survived is a characteristic landscape of scattered mining settlements like the isolated Moor Row and the planned settlement of Cleator Moor, built on previously empty territory. The Jurassic ores of the North York Moors were mined in Rosedale, where the construction of the North Eastern Railway's branch line in 1861 enabled the ore to be taken to Teesside for smelting. The ore was calcined close to the mines in banks of kilns closed at the front by massive iron shutters. Much of the masonry and brickwork of these survive, together with the remains of the complex rail network needed to charge and empty the kilns. The population of Rosedale increased from 548 in 1851 to 2,839 in 1871, and several rows of houses were built. Other miners lodged in the village of Rosedale Abbey, which reverted to its pastoral quiet after the mines closed in the 1880s.[17] Port Mulgrave on the Yorkshire coast was especially built for the shipment of iron ore from other mines on the Moors to Jarrow on Tyneside.

Metallic iron was first smelted in a bloomery, similar to the lead bole hill, and can also be identified from slag scatters. Some were later adapted to make use of water-powered bellows, but had largely disappeared by 1700, by which time charcoal-fired blast furnaces had been in use for 200 years. These were located near good water-power sources for driving the bellows and also near charcoal supplies, which restricted their distribution mainly to forested areas such as the Forest of Dean, the Lake District and the Weald of Kent and Sussex. The furnaces were solidly constructed, about 20 feet high, with arches in two or three sides for the tuyeres which took the blast into the furnace and for tapping the molten iron into a sand-filled pig bed. The fourth side was either built into a bank or had a bridge to the top to enable the furnace to be loaded or charged. Adjacent, too, were a wheelpit and bellows room, together with extensive charcoal stores. The layout of a typical furnace is well demonstrated at Bonawe in Strathclyde Region, where the furnace smelted iron ore shipped in from the Lake District to take advantage of local charcoal supplies. It operated from the 1750s until about 1876 and is now in the Guardianship of Historic Scotland (Plate 17). Several furnaces in the southern Lake District made use of the dwindling charcoal supplies during the eighteenth century. The best preserved is Duddon, with a charging bridge to the furnace mouth. Like Bonawe, this continued operation into the mid-nineteenth century and there are vast charcoal stores at both sites. The demand for artillery during the Napoleonic Wars led to some furnaces being utilised for cannon manufacture and at Rockley, in South Yorkshire, a stone-lined casting pit cut through early eighteenth-century pig beds has been excavated. Several of these have also been discovered in the Weald, particularly adjacent to the furnace at Pippingford.[18]

The introduction, firstly, of coke for smelting and, secondly, of steam power to drive bellows freed the iron industry from the constraints of wood and water. The earliest coke furnaces in Coalbrookdale continued to use water-power, albeit with a steam engine for returning water above the wheel, but by the end of the century large blowing engines were common. It is remarkable that Abraham Darby's experimental furnace of 1709 still survives, but, like the shells of the Bedlam furnaces beside the River Severn, it is far removed from the fiery monsters depicted in the famous de Loutherburg painting of Coalbrookdale by night. Two stone-built furnaces at Morley Park, dating from the 1780s, heralded the spectacular growth of the nineteenth-century iron industry in Derbyshire. These were built by the Hurt family of Alderwasley who already operated forges and were probably the first to use coke in Derbyshire.[19] The furnaces now stand isolated, devoid of their context, in a recently opencasted environment, whereas the brick-built furnace at Moira to the south, dating from 1806, survives almost

Plate 17 The Bonawe charcoal-fuelled blast furnace was established near the shore of Loch Etive in Scotland in 1752–3 by a company of ironmasters from Furness in Cumbria. Special storage sheds were erected for both ore and charcoal, one of which can be seen in the background. The furnace bellows were in front of the left-hand arch, while the iron was tapped from the right-hand arch. This self-contained and isolated industrial community continued to operate intermittently until 1876, very late for a charcoal-fuelled furnace (Photograph reproduced by courtesy of the Royal Commission on the Ancient and Historical Monuments of Scotland).

complete beside the bed of the former Ashby Canal. This furnace was conceived by the impecunious Earl of Moira as a means of redeeming his debts, but was never successful and remained in blast for less than two years.

In the course of the nineteenth century, furnace sizes were increased and several made use of one blowing engine, the result being an industry operating on a far greater scale than previously. Coal was coked in ovens on the same site and casting houses built over the pig beds. This landscape can be appreciated in South Wales, particularly at Blaenafon in Gwent, where there are the massive remains of five stone-built blast furnaces together with a water-balance lift which raised raw materials to the charging bank. Closed about 1900, the site is now in the Guardianship of Cadw. In the same county, the furnaces in the Clydach Gorge, close to the Heads of the Valleys road, have also been excavated and consolidated. Two large blowing engine houses survive in South Wales near Merthyr Tydfil, one at Ynysfach and the other at Dowlais: the former is now an interpretation centre for the town's industries, which included the important Cyfarthfa furnaces belonging to the Crawshay family. However, no examples of the metal-clad blast furnaces which dominated the landscape of the iron industry from the mid-nineteenth century survive and our knowledge of these landscapes is limited to photographic and map evidence.

Ironworking

The product of the blast furnace was pig-iron, which could either be re-melted for castings or converted into wrought iron. Until the mass production of steel in the second half of the nineteenth century, wrought iron was in great demand as a structural material and large numbers of forges were built to convert pig into wrought iron. These remained dependent on water-power and charcoal supplies until the introduction of the puddling furnace for wrought iron at the end of the eighteenth century. This was a coal-burning reverberatory furnace and so all the components of the iron industry could now be situated on the coalfields rather than in its previous sylvan setting. The valleys of South Wales became centres for integrated works, incorporating furnaces for pig- and wrought iron which were then formed into rods and sheets by slitting and rolling mills (Figure 5). Steel replaced wrought iron once it could be made in sufficient quantity: its small-scale production by the cementation and crucible processes was superseded by bulk production in massive Bessemer converters, which have themselves now disappeared, although their scale can be appreciated by the surviving example from Workington which is preserved outside the Kelham Island Museum in Sheffield. The modern steel industry has destroyed most of the early landscape evidence for its own past but the characteristic remains of a cementation furnace at Derwentcote in Tyne and Wear have recently been restored by English Heritage. This is probably the furnace illustrated by the Frenchman Gabriel Jars in his *Voyages métallurgiques* published in 1764.

The water-powered phase of the ironworking industry has left rather more evidence, both on maps and in the field. Hammer ponds are scattered over the Weald of Kent and Sussex and have left distinctive place-name evidence, such as Abinger Hammer, while sites of slitting mills can be found on maps of Staffordshire. The manufacture of agricultural implements and edge tools was widely dispersed and some complete water-powered works survive. The spade mill was particularly common in Northern Ireland, where the National Trust preserve Patterson's spade mill at Templepatrick, while a similar works may be found at Churchill Forge in Hereford and Worcester.

Figure 5 (opposite) The Penydarren works, near Merthyr Tydfil in South Wales, painted in 1817 by Thomas Hornor. He was fascinated by perspective and exploited this technique using light rays from the furnaces in this picture. The works were begun by Francis Homfray in 1784 (Illustration by courtesy of the Elton Collection: Ironbridge Gorge Museum Trust).

At Sticklepath in Devon, an edge tool works with several water-wheels and working forge hammers is now maintained by the National Trust. A more skilled operation was the manufacture of scythes, involving the insertion of a steel cutting edge into a wrought iron blade. Sheffield was the main centre for this and at Abbeydale Industrial Hamlet a working water-powered scythe works is preserved, together with the only surviving crucible steel furnaces which remained in use until the 1940s. The fast-flowing streams in the Sheffield area were lined with forges and grinding shops for the cutlery trade, the best-preserved of which is the Shepherd Wheel on the River Porter. Much of the cutlery itself was made in hand forges attached to houses, operated by the 'little mesters' who dominated the local industry. Wortley Top forge, to the north on the River Don, has been a centre of iron manufacture since the seventeenth century, but from the mid-nineteenth century specialised in the production of railway waggon axles. The Sheffield Trades Historical Society, who maintain this site, are dedicated to preserving the iron and steel working crafts of this most important area.

Hand forges in the Black Country were used for the manufacture of nails and chains rather than edge tools. Nails were made in backyard workshops, often associated with late-nineteenth century terraced housing, such as the row moved from Cradley Heath in the Black Country to the Avoncroft Museum in Hereford and Worcester. Larger workshops, with multiple forges and hammers, were used for the manufacture of chains and some can still be found in the Black Country. At Mushroom Green, the distinctive façade of a chain shop with its multiple chimneys can be appreciated, and chain is still made by hand here by the Black Country Society. A similar example has been moved from Cradley to Avoncroft. Nearby Redditch became an important needle-making centre, and a water-powered needle mill forms the basis of the Forge Mill Museum.

The blacksmith's shop could be found in every village and many examples survive, some associated with wheelwright's shops. An unusual landmark is the conical tower of old horseshoes associated with the village forge at Scarrington in Nottinghamshire, while the forge at Tinwell in Rutland has a horseshoe-shaped porch of stone. Equally common were the rural foundries, recognisable from their melting furnace chimneys and ventilated roofs, which also made and repaired agricultural implements. Many of these survive in East Anglia, such as at Acle and Panxworth, but elsewhere many formed the nucleus of subsequent large-scale engineering works like Garretts at Leiston. A highly specialised foundry trade was the production of bells from alloys of copper and tin. Many temporary furnaces were set up on site to recast bells, but now only two permanent bell foundries remain, one at Whitechapel in London and the other at Loughborough in Leicestershire. At the latter there are several furnace chimneys on the foundry building, together with a bell-testing tower, and the firm of John Taylor has set up a small museum of the industry.

*

Man's quest for metals has taken him to some of the most remote parts of Britain. Leats and reservoirs for water, open cuts for ore, countless mine shafts, some with engine houses and dressing floors, are still features of the landscape. However, current concerns about heavy metal contamination together with funding for derelict land reclamation have led to the destruction of many of these features without any record being made. This work has created some sanitised industrial landscapes where isolated features like ruined engine houses have been surrounded by acres of green grass, totally devoid of their original context. At least increasing interest in mining landscapes has ensured the survival of some structures, as in the comprehensive Mineral Tramways Project in Cornwall which seeks to create a long-distance footpath based on the former mineral railways in the Camborne–Redruth area and link together a series of restored engine houses. Perhaps only in the remoter parts of Wales and the Pennines can the unaltered mining landscape still be appreciated.

5

Clothing the people

The vagaries of the British climate have made clothing a necessity from the very beginning. Man's ingenuity was taxed in order to convert natural and animal fibres into wearing apparel and so successful were his endeavours that clothing also became a means of displaying rank and wealth. The demands of fashion as well as the growth in population meant that the production of fabrics and footwear became two of Britain's major industries from the late Middle Ages onwards. The rise in living standards also led to the use of textiles and leather for house furnishings in the form of upholstery, draperies and carpets. In the landscape, the wealth generated by the industry is revealed in the elaborate 'wool' churches of East Anglia and the elegant clothiers' houses of the West Country. By contrast, the means of production can be identified in the weavers' houses, stockingers' workshops, water- and steam-powered mills and factories which are a feature of the landscape in many parts of Britain.

For centuries men have been wearing animal hides and skins for clothing and footwear. The curing and tanning of leather became a highly organised industry, and boots, shoes and leather for clothing and furnishing were produced. The most commonly used thread for textiles at the beginning of our period was wool from both sheep and goats, ranging in quality from the fine merino used for the new draperies to coarse wool for blankets. Generally, long staple wool was made into worsted cloth, while the many types of 'woollens' made use of short staple fibres. A lighter luxury fabric was made from silk, whose manufacture was begun by the silkworm producing filaments for its cocoon which had to be unwound, twisted and doubled to produce a usable silk thread. The silkworm thrived on a diet of mulberry leaves which grew better in France and Italy than in England, and it was to process imported silk that the first British water-powered factory was built in 1702 at Derby, although this made use of stolen Italian technology. Plants were another source of thread, the woody stems being soaked in water and crushed to release the fibres before they could be spun. Fine linen thread was produced from the flax plant and coarser sacking, hessians and rope twine from hemp and jute, all of which were grown in eighteenth-century Britain, although the raw materials were later brought over from the Baltic states. Early in the century, calico or finished cotton fabric from India was being imported, but the sale of British-made cotton textiles was forbidden by law to protect the interests of both the woollen manufacturers and the East India Company. The law had already been circumvented by the manufacture of fustian, a fabric with a linen warp and cotton weft, but collapsed in the late eighteenth century when the fluffy cotton seedheads or bolls began to be imported in large quantities from America. The conversion of these into thread inaugurated the mechanisation of textile production which had a dramatic effect on the lifestyle of thousands of people as well as the addition of large numbers of mills, factories and industrial communities to the landscape. In this chapter the processes of textile manufacture will be discussed in so far as they affected the buildings and landscape, followed by the less attractive tanning industry and then the manufacture of leather products.

The animal and plant fibres had first to be cleaned and then aligned, either by carding for short fibres or by combing for the longer fibres. These were then spun into thread before weaving or knitting into fabric, following which various finishing processes were necessary according to the type of thread. Woollen fabric was the most complex, requiring fulling and stretching or tentering before being dyed. Patterned fabrics could be produced by weaving different coloured threads, using complex looms, or later by block printing plain cotton cloth. The marketing system for finished cloth varied from the 'piece halls' used by the yeomen clothiers of Yorkshire to the warehouses used by the capitalist merchant clothiers of south-west England and the hosiers and lace merchants of the East Midlands. These different forms of industrial organisation are reflected in the built environment of the main textile-producing regions of Britain, each of which will be considered in turn.

The woollen industry

Hand spinning of wool was once a universal domestic secondary employment as can be seen from the equipment listed in probate inventories. This was often carried on out of doors to get maximum light, but where shelter was needed, as in north-west England, spinners sat in special galleries built out on the first floor of the house. Examples of these may be found in Sedbergh and Ravenstonedale as well as in parts of the Lake District. Hand weaving, on the other hand, was generally concentrated in specialist textile districts such as East Anglia, south-west England and the south

Pennine area of Yorkshire and Lancashire, spun thread being imported from other regions to boost local supplies.

Probably the most evocative landscape of the woollen industry is the Saddleworth area of Greater Manchester, formerly part of the West Riding of Yorkshire. On marginal land, farmers occupying ever-decreasing holdings due to the division by inheritance of estates between whole families were forced to supplement their income by making cloth. Here can be seen the remains of all periods of textile production, from the medieval upland spinning and weaving industry to water-powered mills in the valleys, purpose-built domestic loomshops and then large mills; these were first on sites beside canals and then served by railways along which coal supplies came to fuel their steam engines. Terraces of workers' houses contrast with elaborate mill-owners' mansions, while Victorian town halls, churches, schools, dyeing sheds and works for producing textile machines complete this panorama of the woollen industry. In the early eighteenth century, Defoe, travelling from Rochdale to Halifax, remarked on the landscape:

> we could see that almost at every house there was a tenter, and almost on every tenter a piece of cloth, or kersey, or shalloon, for they are the three articles of that country's labour; from which the sun glancing, and, as I may say, shining to us, I thought it was the most agreeable sight that I ever saw.[1]

Individual weavers placed their looms in upstairs rooms, or in buildings attached to laithe houses, as at Peckett Well north of Hebden Bridge. The laithe houses, common throughout the Yorkshire Dales, combined within one building accommodation for both men and animals, separated from one another by a passage with waggon access through double doors. A wealthy farmer might set up as a yeoman clothier, supplying thread to weavers and marketing the finished cloth. At New Tame and Harrop Green, the main farmhouse is the nucleus of a cluster of cottages which have loomshops with multiple windows and taking-in doors on the upper floors (Plate 18). Larger looms incorporating the flying shuttle required more floor area and rows of stone cottages were built, such as those at Heights, Butterhouses and High Kinders, with upper floor loomshops; all have long windows at front and rear and the latter exterior steps to the upper floors. Improvements in spinning machinery, such as the jenny, also required more floor area, and the subsequent mechanisation of the spinning process ensured continuous employment for these cottage weavers well into the nineteenth century. They resisted the introduction of the power loom in the first decade of the nineteenth century, the Saddleworth merchants and master clothiers reaching an agreement that no home should contain more than 160 spindles and five looms. The factory system was kept at bay for several more decades.

Plate 18 New Tame Fold, near Saddleworth, Greater Manchester, formerly in the West Riding of Yorkshire. A community of Millstone Grit buildings with former domestic workshops grouped around a farmhouse. The yeoman clothier brought in spun wool to be woven into cloth, which was then taken to the cloth halls for disposal. Many of the buildings have long multi-paned windows to give light to the hand looms. The three-storey building on the right has long windows to three floors and a taking-in door for materials in the end wall.

Since the twelfth century, water-power has been used for fulling woven woollen cloth which was pummelled in liquid in troughs to compact the fibres and also to remove the natural oils and the size applied during manufacture. The cloth was then 'tentered' or stretched on hooks on posts in the open air to dry and restore its shape: rows of stone tenter posts can be seen at Wall Hill in Dobcross, while yarn and cloth were often spread to dry on specially built stone walls with a stepped side facing south. The scribbling machine was developed from machinery produced for cotton preparation and used for breaking down the wool fibres preparatory to spinning; this machine was water-powered and many were added to existing fulling mills. At Delph, the three-storey Shore Mill survives to indicate the small scale of these early water-powered mills which supplied local spinners and weavers. Spinning was the next branch of the industry to succumb to water-power and spinning mills housing multi-spindle mules were built in the river valleys. The River Tame in Saddleworth was no exception and the remains of ponds and leats for water-power can be seen at Royal George Mills, Frenches Mill and Giles Mill; at the confluence of the Diggle Brook is the wheelpit of the Mytholme Mill. The Brownhill Bridge Mill standing on the Diggle Brook tells the story of many mill sites, the three-storey stone building having been erected as a fulling and scribbling mill in the 1770s, but it was later used for spinning and ended its industrial career as a tannery.[2]

The landscape of steam power is perhaps more obvious in Uppermill, where stone-built multi-storey mills with tall chimneys and water-towers were grouped first along the Huddersfield Narrow Canal and then along the railway which bought out the canal company. The canal, which snakes its steep descent from the summit tunnel at Standedge down to the Ashton Canal and Manchester, was opened in 1811. The Huddersfield and Manchester Railway followed the same route in 1844, crossing the canal and the river valley by a high curved stone viaduct. This transport corridor encouraged a new phase of mill-building, and several steam spinning mills in Uppermill date from 1835 to 1880. At Dobcross, a large factory manufacturing power looms was established in 1861 beside the canal with a branch line from the railway; the extensive top-lit machine shops and fine office building are testament to their one-time prosperity. The later looms which they made became too heavy for multi-storey buildings and the final phase of mill development is reflected in the single-storey weaving sheds with their saw-toothed, north-light roofs, often concealed by high walls, their presence betrayed by roundel windows or vents and downpipes from the roof valley gutters.

Cloth from Saddleworth was originally taken to market by packhorse. The first Yorkshire cloth hall was established in 1545 at Heptonstall, on the east–west hilltop route from Halifax to Burnley. Here merchants bought finished cloth and the yeoman clothiers more thread for weaving. Heptonstall is one of several hillside hand-weaving villages with an associated river valley settlement. Below Heptonstall, the new village of Hebden Bridge was established when first the Rochdale Canal and then the railway provided a trans-Pennine route for coal and goods. Here, Nutclough Mill, established in the 1850s and later occupied by a co-operative for fustian manufacture, illustrates the phases of mill development with its multi-storey spinning mill and single-storey north-light weaving sheds on the valley floor. Terraces of earlier weavers' houses and later mill-workers' houses are perched precariously alongside the road to Keighley. Fustian and corduroy manufacture still continues at the small Pecket Well Mill, north along this road. Dated 1858, the stone-built warehouse block, with an unusual integral chimney, has weaving sheds attached and still provides employment for an old-established hillside weaving community.[3]

Similar hillside and valley settlements can be seen in the Colne Valley, the loomshops of villages like Golcar looking down upon the large water- and then steam-powered mills which line the river and canal banks through Marsden, Slaithwaite, Linthwaite and Milnsbridge (Plate 19). These relied upon Huddersfield as a centre, while Halifax served the Calder Valley mills at Hebden Bridge, Mytholmroyd, Luddenden and Sowerby Bridge. The splendid Halifax Piece Hall, opened in 1779 as a cloth market, has been restored and is partly occupied by a museum to the domestic woollen industry. The Dean Clough Mill complex, also in Halifax, once gave employment to 5,000 people making carpets and carpet loom machinery; a range of six- and seven-storey mill buildings now houses an industrial park. Bradford grew rapidly during the early nineteenth century and became the marketing centre for worsted cloth. The warehouse quarter, known as 'Little Germany' from the early influx of immigrant merchants, is an area of multi-storey stone-built warehouses whose ornate style was intended to attract customers for finished cloth. Much of the raw material by this time was sold through the ornate Gothic Wool Exchange of 1867. Two progressive entrepreneurs established large mills in the Bradford area, the first being Titus Salt who built Saltaire between 1853 and 1868 to process mixtures using worsted, alpaca and angora hair. His mills straddle the Leeds and Liverpool Canal, the main chimney modelled on a Venetian campanile. The complex is surrounded by a complete village settlement and the quality of the public buildings and houses echoes that of the mills themselves. The other large mill at Manningham, also architecturally pretentious, possesses an even taller chimney but plainer houses; this was built in the 1870s for Lister & Co. for silk manufacture, but later produced velvet. These two mills contrast sharply with the scale of the domestic woollen industry of Saddleworth and illustrate the rapid transition of the industry in less than a century. The landscape of the Pennine woollen industry is one of the richest areas of Britain's industrial heritage.

The Yorkshire woollen industry prospered at the expense of that of East Anglia and south-west England. The former, which helped to make Norwich the second city in England in the sixteenth century, had all but died out by 1700, leaving a rich legacy of wool churches, such as Long Melford, and timber-framed clothiers' houses, as in Lavenham. An attempt was made to revive the worsted industry in Norwich in the mid-nineteenth century, leaving the gaunt red-brick mill with its domed stair tower beside the River Wensum, now occupied by a printing company. The industry in the West Country, however, survived much longer. Once dependent upon heavy broad cloth, it had diversified into the lighter, high quality fabrics which were fashionable in the seventeenth century. While the Pennine industry was dominated by the independent yeoman clothier, that of the south-west was ruled by wealthy merchant clothiers whose elegant stone-built houses grace the streets of towns such as Trowbridge and Bradford-upon-Avon. Many can also be found alongside mills in the Gloucestershire valleys, the clothiers not being too proud to live next door to their works. At Lodgemore near Stroud, the elegant stone-built eighteenth-century house, now used as company offices, contrasts with the red- and blue-brick Victorian mill which replaced the earlier stone one. The West Country industry was organised on a putting-out basis, but there is little obvious evidence of domestic weaving workshops like those of the Pennines,

Plate 19 Disused woollen mills on Tanyard Lane at Milnsbridge, near Huddersfield, West Yorkshire. Many new woollen or worsted mills were erected here once the Huddersfield Canal and later the Huddersfield and Manchester Railway opened communications along the Colne Valley. A fine stone-built multi-storey mill stands on the right, with later north-light weaving sheds on the left. In the background are later steam-powered mills complete with towers housing water storage tanks for their fire sprinkler systems.

although the attractive terraces of Tory and Middle Rank in Bradford-upon-Avon were occupied by cloth-workers. It is possible that the small, four-gabled houses in Nailsworth and the Stroud Valley once contained broad looms in their attics. The principal product of this area was Stroud scarlet for uniforms, and contemporary paintings show the hillsides draped with red cloth on tenters.

The most important elements in the textile landscape of this area are the stone-built mills, water-powered to begin with and then steam-powered once canals enabled coal to be brought in. The rivers flowing west from the Cotswolds were lined with cloth mills: along the Frome and its tributaries can be seen St Mary's Mills at Chalford, still with both water-wheel and steam engine, the massive Dunkirk Mill at Freshford, now converted to luxury housing, Ebley Mill in Stroud, newly adapted for council offices, and the finest of them all, the brick-built Stanley Mill with stone quoins and Venetian windows giving a view of the impressive interior ironwork dated 1813. The competition from Yorkshire and collapse of export markets led to the decline of the woollen industry in this area early in the nineteenth century, and by 1850 many of the mills were already converted to alternative uses, one of which was the manufacture of walking-sticks and canes so fashionable in Victorian England. The River Avon also powered a series of impressive mills, notably the brick-built Abbey Mill in Malmesbury and the similarly named but stone-built mill near the river bridge in Bradford-upon-Avon. Below Wootton-under-Edge, the elegant Kingswood Mill, once powered by the Little Avon, has been converted to offices but retains its circular wool drying stove, similar to those which can also be found in Melksham and Bradford-upon-Avon. Another unusual building associated with the woollen industry is the ventilated handle house for drying teasels, such as that which survives at Studley Mill in Trowbridge straddling the River Biss. Heads from the fullers' teasel were used for raising the nap of woollen cloth, both in the hand process and later in powered gig mills. The number consumed was enormous: the sixty gig mills operating in Saddleworth in the 1830s used 21 million teasel heads a year. These were cultivated in various parts of England, but many were imported from Europe.

In Wales, several small water-powered woollen mills continue to produce items for the tourist trade, perpetuating an earlier phase of an industry which survived longest in the Teifi Valley in Dyfed. Here, flannel was produced for the coalminers of South Wales by water-powered machinery until well into the twentieth century, and the small scale of production can still be appreciated in mills such as Melin Teifi and Rock Mill near Drefach Velindre.[4] Further north, in the upper Severn Valley at Llanidloes and Newtown, surviving weavers' windows indicate the former presence of the once important woollen manufacture of this area. Woollen manufacture was also important in Scotland, and hand weaving still survives in the Hebrides. In the Hillfoots area of the Central Region, a fine series of water-powered mills was built in towns below the Ochil Hills, notably the Clock Mill in Tillicoultry, now a museum, and Strude Mill in Alva, now converted into flats.[5] But the single most important manufacturing industry in eighteenth-century Scotland, as in Ireland, was that of linen, which was also made on a small scale over much of England. Boards of trustees were set up in both Ireland and Scotland to promote the linen industry, which flourished and survived the competition from cotton later in the century.[6]

The linen industry

Spinning and weaving of linen thread were carried out on a domestic basis and the only processes occurring outside the home were scutching, or breaking up the flax stems in water-powered mills, and bleaching. Many scutching mills were built in the central lowlands of Scotland but few survive in recognisable form today. However, in Ulster alone there were over a 1,000 such mills in the middle decades of the nineteenth century and many can still be traced today in the Counties of Antrim, Armagh and Down.[7] Water-powered flax heckling, similar to carding in the woollen industry, and spinning were introduced in the late eighteenth century, followed by steam operation in the 1820s, but the weaving process was not powered until the 1840s. Loomshops for linen weaving, unlike those for wool, were often at cellar and ground-floor level to promote the damp atmosphere necessary for the brittle linen thread; many continued in use in the first half of the nineteenth century and examples can be found in Scotland, Lancashire and North Yorkshire.[8] Many mills originally constructed for powered spinning later became integrated complexes with the addition first of hand loomshops and then powered weaving sheds, as in the woollen industry, and finally fully integrated steam-powered mills were built.

The woven fabric was bleached by spreading it out in sunlight on the fields surrounding the crofts: the market demand was for white linen and so bleaching was a particularly important part of manufacture. Near large mills, public bleachfields were laid out. These can be identified from map evidence although they have mostly been built over. The long lengths of cloth laid out in rows, often with fabric on tenters nearby, must have presented a spectacular sight and the bleachfields were frequently guarded by watchmen to prevent theft. Chemical bleaching, pioneered in Glasgow in the early part of the nineteenth century, replaced field bleaching and became a factory process which was a great deal faster than the old methods. After bleaching, the final stages in finishing linen involved beetling, or pounding with heavy wooden hammers, and calendering, or rolling between cylinders, both of which gave a smooth finish to the cloth. The former was usually carried out in water-powered mills, several of which survive in Northern Ireland, whereas the calendering process was occasionally undertaken by bleachworks until specialist calendering mills were developed. The 'Belfast' roof was developed at the end of the nineteenth century to cover bleach- and ropeworks as well as other trades requiring a lightly constructed roof over a large floor area. Its distinctive shallow convex shape is due to the timber lattice trusses covered by wood and felt sheeting, sometimes having glazing panels or clerestory ventilation louvres.

In Scotland, fine linen had been produced around Glasgow in the west until it was superseded by cotton, while the east, apart from Dunfermline, specialised in coarser fabrics. The rising price of imported flax from Baltic countries caused many mills to change to the production of jute and hemp, also using imported raw materials. Dundee became known as 'Juteopolis', the number of its mills increasing from five in 1800 to sixty in 1864. By then, it contained both the world's biggest linen and jute works.[9] Many linen mills were converted to jute production, for example, the Logie Works, founded in 1828, had long four- and three-storey blocks and a staircase tower. At Dens Works on Princes Street, the spinning mill with its domed bellcote has been converted into flats, following the demolition of most of the complex. But the townscape of Dundee, despite the collapse of the industry, is dominated by purpose-built jute works, particularly the 280-feet high red- and white-brick chimney stack of the Camperdown Works of 1861–8. Not all the buildings survive here either, but the impressive four-storey cast iron-framed High Mill with its bellcote has been converted for other purposes. The pedimented Tay Works along the Lochee Road are of a similar date and have been adapted to student and office accommodation.

Fine linen thread production continued in Northern Ireland, and by 1900 around 80 per cent of the British flax-spinning capacity was located there. Large multi-storey steam-powered spinning mills became a major feature of the Belfast townscape. In England, linen manufacture was widespread but generally on a smaller scale. Even so, individual manufacturers pioneered new techniques in fireproof cast iron-framed buildings. Bage's flax mill in Shrewsbury, later converted to maltings, was the first of these buildings in 1797, to be followed by others in Leeds built by Marshall and Benyon. Between 1838 and 1841 Marshall added to his existing complex the far more pretentious Egyptian-style Temple Works, modelled on the Temple of Karnak. This was an integrated textile complex, including a vast single-storey building lit by glass domes in a flat roof. Linen manufacture flourished elsewhere in the Yorkshire woollen area, with important centres at Knaresborough, where relics of the domestic industry survive, and at Pateley Bridge where the massive Glasshouses flax-spinning mill of the early nineteenth century contrasts with the compact water-powered mill with a 35-feet diameter external wheel at Foster Beck, built as late as 1887. In southern England, rising demand during the Napoleonic Wars stimulated both the linen and hemp industries for the production of sailcloth and ropes. In Dorset and south Somerset rope walks were often attached to flax and hemp mills, as can be seen at Castle Cary. In Bridport, still an important rope- and net-making centre, only the grid street layout betrays the former presence of numerous open rope walks, although many of the associated stone-built multi-storeyed warehouses still remain. Nets, sheets and ropes also had important agricultural uses and consequently their manufacture was not confined to coastal areas. Most large towns had rope walks, often open to the air and surrounded by terraces of houses. Elsewhere, centres developed to serve particular local industries, as at Hailsham in East Sussex where the firm of Thomas Burfield was established in 1785 to make cordage for the hop gardens as well as sailcloth for windmills. Nevertheless, the largest ropes were needed for shipping and many naval dockyards had their own rope walks. Some of these, like the 1,140 feet long double rope walk at Chatham, constructed for the Navy during the 1780s, were the longest single industrial buildings yet seen in Britain.

The cotton industry

Flax produced a stronger thread than cotton and had been used for warps in fustian weaving, based on the Lancashire slopes of the Pennines from the early eighteenth century. Technological improvements later in the century, combined with changes in the law, enabled pure cotton cloth to be made. As a new industry, it was more receptive to the introduction of powered machinery than the long-established woollen industry and pioneered organised factory production. Strangely enough, this system developed not in Lancashire, as might have been expected, but in the East Midlands, where the demands of the hosiery industry provided a ready market for cotton thread and there was less resistance to technological innovation. Both James Hargreaves and Richard Arkwright developed their spinning machines in Lancashire but first put them to actual use in Nottingham. It was Arkwright's water frame which successfully produced a cotton warp thread, and his move to Cromford in Derbyshire led to the establishment of the first water-powered cotton-spinning community in 1771. His Cromford Mill was the first of many which followed a specific pattern, two or more storeys high, about 70 feet long and 30 feet wide to allow maximum light on to the machines. Power from the waterwheel was transmitted to each floor by a vertical shaft with gearing to line shafting, which took the drive to the machines by belts. Over 200 different mills from the Midlands to central Scotland were licensed to use the Arkwright machine.[10] Several survive, including the lower storeys of the original mill at Cromford, and characteristic buildings can still be seen at Wirksworth and Darley Abbey in Derbyshire and Rocester in Staffordshire. Jedediah Strutt built a series of mills in Belper from 1776, and his association with Charles Bage of Shrewsbury led to his adoption of cast iron framing and fireproof construction for his North Mill, built 1803–4, which still stands. This important structure retains the pit for the 23-feet diameter wheel, while the adjacent crescent weir on the River Derwent once stored water for this and other mill wheels in Belper. North Mill is now dwarfed by the massive red-brick steam-powered East Mill which was erected in 1912.

The progression from hand- to water-powered spinning entailed a change in settlement patterns, the workforce becoming concentrated around the power source rather than living in scattered communities. They also had to submit to a regular routine in a factory environment rather than controlling their own pace of output. In the Derwent Valley of Derbyshire, the new cotton workers' communities were often grafted on to existing villages where other crafts had previously dominated, like lead-mining in Cromford and nail-making in Belper. In the former village, Richard Arkwright constructed stone terraces of houses with workshops for weaving on the upper floors, whereas in Belper, the Strutts promoted the nailing industry: both provided employment for men whose wives and children were employed in the mills. They also took steps to feed their workforce by the provision of fresh milk and produce from their own farms. Further south at Darley Abbey, the Evans family provided several squares of three-storey cottages to house the workforce for their Boar's Head Mills on the east bank of the Derwent.[11]

In Scotland, the Falls of Clyde were harnessed for cotton spinning in 1785 through the efforts of David Dale, a local cloth merchant, in conjunction with Richard Arkwright. In ten years they constructed mills at New Lanark employing nearly 1,500 people, who were housed in long rows of tenements on the Scottish pattern. Dale's daughter married Robert Owen, who became manager of New Lanark in 1800 and was able to put into practice his theories concerning the improvement of character through the environment. The Institute for the Formation of Character and the massive school are evidence of his efforts.[12] The isolation of the community has resulted in its survival almost in entirety, as is the case at Stanley, near Perth, a lesser known water-powered cotton-spinning community. The red-brick mills on this attractive site beside the River Tay date originally from 1785 and the nearby village contained 500 inhabitants by 1823.

Further south, Samuel Greg harnessed the River Bollin in Cheshire to provide power for a new red-brick mill at Styal in 1784. He built his own house adjacent to the mill but sited his factory colony half a mile away: this consisted of several rows of good quality housing and an apprentice house. He was proud of his development and welcomed observers to both the mill and the village, a practice continued by the National Trust who now manage what is probably the only working water-powered cotton mill in the world (Plate 20). At Papplewick in Nottinghamshire, a Scottish bleaching family, the Robinsons, established a series of spinning mills with associated housing on the River Leen. Experiencing severe problems over water supply, they installed a rotative steam engine in 1786 and so inaugurated a new era in the cotton industry.[13] Although most of these early cotton-spinning communities survived by adding steam engines to supplement water-power at their mills, the future of the industry now lay on the coalfields. Lancashire, with Liverpool as the main port for cotton imports, was able to regain its former pre-eminence in the textile industry.

Arkwright-type cotton-spinning mills had been built in both Lancashire and Yorkshire and many, as in the East Midlands, continued to operate by adding steam engines. By the beginning of the nineteenth century, the larger spinning mules required more floor space and mills increased in size. Their expansion in width was limited until the advent of gas lighting reduced the dependence on natural light, and their length was restricted by the stresses imposed on the horizontal driving shafts on each floor. Beam engines were either installed within the mill building or in separate engine houses, with adjoining boiler houses and tall chimneys; water, of course, was necessary to supply the boilers and both reservoirs and condenser ponds still formed part of the landscape. In Lancashire, these were known as lodges and in Preston alone in 1889 there were ninety-seven lodges, some even licensed to collect surface water from the town streets.[14] These

Plate 20 The long range of brick buildings at Quarry Bank Mill, Styal, in Cheshire on the River Bollin. The mill was founded in 1784 by Samuel Greg for cotton spinning and subsequently extended for weaving too. The village and apprentice house built by the Gregs for their workers stand some distance to the north of the mill. The complex is now owned by the National Trust.

early nineteenth-century mills were built for spinning since, as in the woollen industry, weaving was carried out by hand until the 1840s, often in ground-floor or cellar loomshops which provided the humidity necessary for cotton as well as linen. The heavy powered looms which were then introduced resulted in the construction of single-storey north-light sheds either attached to an existing mill or forming a totally separate establishment, which included a warehouse for the spun thread. Several of these large spinning mills can still be seen in an attractive setting alongside the Macclesfield Canal at Bollington in Cheshire, while Frostholme Mill near Todmorden is a typical weaving establishment with its single-storey shed and multi-storey warehouse. Queen Street Mill, in the industrial village of Harle Syke on the edge of Burnley, was built for weaving in 1894 and little of its machinery has changed since: it has now been preserved as the last steam-powered cotton-weaving mill in Britain.

The end of the cotton famine which had been caused by the American Civil War in the 1860s initiated a new era of mill-building, financed by speculative capital rather than by the master spinners. Rather than being strictly functional, these new larger mills exhibited some architectural pretension, usually in red brick with terracotta decoration. Many were designed by specialist mill architects such as the Stotts and the Bradshaw and Gass partnership, whose structural innovations enabled wider pillar spacings and hence larger machines to be accommodated.[15] The advent both of the horizontal mill engine and rope drive by the 1880s altered mill layouts. In the rope transmission system, the horizontal drive shafts on each floor were turned by pulleys driven by cotton ropes connected to the massive engine flywheel. The mill engines were contained in external engine houses, which were sometimes located on the end of a mill block whose blank wall indicated the intention of adding another block to create a double mill. Fire was an ever-present hazard in cotton mills in spite of fireproof construction and flat roofs replaced gabled roofs so that they could be utilised as water storage reservoirs. The introduction of pressure water sprinklers in the 1880s added the towers which are such a distinctive feature of most Lancashire cotton mills, many carrying the mill name emblazoned in large letters. Chimney stacks were an even more prominent feature and were consequently often ornamented, for example, the Venetian campanile which dominates the India Mill in Darwen. Nowhere can the scale of the cotton industry be better appreciated than in Oldham, where Victorian and Edwardian mills remain a dominant feature of the landscape despite their reuse for other purposes. For nearly 100 years from the middle of the nineteenth century, Oldham had greater cotton-spinning capacity than anywhere else in the world apart from the USA. There were 300 mills in the town by 1900, some with 100,000 spindles.[16] These uncompromising structures were four or five storeys in height and square in plan, their red-brick walls punctuated by regular rows of tall windows (Figure 6). They dominated the infill of rows of terraced houses, creating a landscape made familiar through the paintings of L.S. Lowry.

The silk industry

Lancashire replaced the East Midlands as the centre for cotton production but the midland hosiery industry continued to provide a ready market for thread. Silk and worsted dominated the industry until the end of the eighteenth century when it became possible to knit cotton hose. The first machine-made knitted hose were of silk, and it was to satisfy this demand that the first successful water-powered silk mill was set up in Derby by Thomas Lombe in 1718. This precursor of the factory system was noted by Daniel Defoe in 1725:

> Here is a curiosity in trade worth observing, as being the only one of the kind in England, namely, a throwing or throwster's mill, which performs by a wheel turn'd by the water; and though it cannot perform the doubling part of a throwster's work, which can only be done by a hand wheel, yet it turns the other work, and performs the labour of many hands.[17]

Figure 6 An advertisement for Lees & Wrigley's Greenbank Mills in Oldham, Lancashire, *c.* 1910. It illustrates the juxtaposition of terraced houses for the workforce with multi-storey steam-powered cotton-spinning mills having north-lit sheds behind. The first of these mill blocks was built in 1816–17 by John Lees (Illustration by courtesy of *Oldham Evening Chronicle*).

The hand silk industry had been centred in the Spitalfields district of London and in Suffolk, following the influx of Huguenot refugees from France at the end of the seventeenth century. Weaving looms were operated in attic workshops, some of which can still be seen in Fournier Street in Spitalfields, while small workshops developed in the Suffolk towns of Haverhill and Sudbury. Lombe's Derby Mill was the forerunner of many others, both in Derby itself and in isolated villages with good water-power sites. A particularly good example of an industrial community is the mill and associated housing at Maythorne near Southwell in Nottinghamshire, dating from the early nineteenth century. Even earlier in origin are the Galgate silk mills near Lancaster, begun in 1792 with additional buildings up to 1851 – terraces of workers' houses survive nearby. The silk taken by barristers upon their nomination as QCs is still produced at a fine red-brick silk-weaving mill dating from 1815 on the River Test at Whitchurch in Hampshire.

The introduction of steam power for silk throwing enabled larger mills to be built, such as the gaunt Rykneld Mill in Derby and the more attractive pedimented mills to be found in several east Cheshire towns, notably Congleton and Macclesfield, together with Leek in Staffordshire. Macclesfield had twelve silk-throwing mills by 1817, weaving being carried out in the garrets which are still a feature of many terraced houses in the town, such as Paradise Street. The introduction of the jacquard loom in the 1820s necessitated raising the roof of the weaving garret to accommodate the additional height required for the punched cards which determined the pattern in the fabric.[18] A similar change was necessary in the mills once power weaving was introduced. The massive Sunday School in Macclesfield, built for silk-weavers' children in 1813, and Paradise Mill now provide an interpretation centre for the silk industry. Another centre for silk production was Coventry, which specialised in fancy ribbons. These required narrow jacquard looms and the weaving garrets still to be found in Coventry are tall with large deep windows. An interesting development in the town was the cottage factory, where rows of cottage workshops were built around a common central engine house. Beginning in the late 1840s, these prolonged the domestic system, allowing power looms to be used in the home. Most have disappeared, but some parts of Cash's 'hundred' houses, alongside the Coventry Canal, have been converted into flats and their large top-floor windows are still obvious.[19] The British silk industry declined following a free trade treaty with France in 1860, after which duty-free French silks flooded the market.

Hosiery and lace

Silk hosiery had satisfied the luxury market, but with rising population during the eighteenth century first worsted, and then cotton, became more important than silk for the production of knitted hose and fabrics. The East Midlands were producing more than 90 per cent of the output by 1800, all of it on hand knitting frames in domestic workshops.[20] When the industry moved here from London in the late seventeenth century, knitters inserted long windows into their homes to allow maximum light to fall on the complex knitting machine. These can be recognised in several timber-framed buildings in the region. As the industry grew in the eighteenth century, speculative builders began to take advantage of the knitters' need for houses containing a workshop large enough to contain their frames. They provided rows of houses for rent incorporating workshops with large windows: if the windows were placed on the first or second storey, the light was better than on the ground floor but strengthening was needed to support the weight of the knitting frame. Houses of this kind continued to be built until well into the nineteenth century, as an example dated 1857 in the brickwork on the gable end of a cottage in Calverton in Nottinghamshire demonstrates. As frames became larger, small one- or two-storey workshops with windows on several sides were built to house them. Two good examples have become museums of the industry at Ruddington in Nottinghamshire and Bushloe End at Wigston near Leicester.

It was not until the 1870s that the hosiery industry adopted powered machines, the last of the textile industries to do so. The warehouses from which yarn was distributed to the domestic knitters were converted into factories with the addition of steam or gas engines. Knitted garments were now being produced as well as hosiery, and the last three decades of the nineteenth century witnessed the building of a large number of hosiery and knitwear factories. The sudden urbanisation of the industry resulted in rapid town expansion, with row upon row of well-built terraced brick houses being constructed. Although much has been swept away in mid-twentieth-century town improvement schemes, the late nineteenth-century townscape can often be reconstructed from the Goad's Fire Plans which were large-scale maps made for insurance purposes from the 1880s until recently. Where these exist, they indicate the number of storeys on factories, the materials from which they were constructed and occasionally the source of power and position of boilers. The 1892 plans for Leicester show the canyon-like streets which were developed, lined by powered hosiery factories with terraced housing crammed in between. This pattern was repeated in many East Midland towns and villages. The scale of change can perhaps be best appreciated in Hinckley by comparing the seventeenth-century timber-framed domestic workshop on Lower Bond Street with the red-brick multi-storey factory opposite built between 1877 and 1910.[21]

The similar transformation of Nottingham was based partly on hosiery but also on machine-made lace production which dominated the nineteenth-century town as lace became more fashionable for Victorian clothing and household fabrics. The lace frame developed from the hand stocking frame in the late eighteenth century, and purpose-built housing incorporating workshops characterised the early years of the lace industry much as it did hosiery, as can be seen in surviving three-storey houses with topshops in Stapleford, west of Nottingham. Inventions by Heathcoat and Leavers early in the nineteenth century resulted in the production of lace machines which were considerably larger than hosiery frames. These were installed in workshops by the 1830s and steam power soon followed, so that by 1850 there were few hand lace frames remaining. A feature of the lace industry was the tenement factory, often built in the satellite towns and villages of Nottingham where land was cheaper. These enabled small producers to rent floor-space and share a common power source, the factories themselves often being built by speculators or even lace-machine makers. The tenement factories were generally on a similar scale to the cotton mills of Oldham and far larger than most hosiery factories in the region. As in other textile industries, late nineteenth-century machinery became so heavy that multi-storey buildings were no longer suitable and so single-storey factories intruded on the urban scene. Examples of all phases of lace factory development may be seen in Long Eaton, lining both banks of the Erewash Canal.[22]

The early lace machines produced a plain net fabric which could be embroidered or decorated with handmade lace motifs. Following labour problems in the East Midlands in the second decade of the nineteenth century, some lace manufacturers opened new premises in areas of south-west England where the lace motifs were being made and female labour was available to embellish the plain net. John Heathcoat, inventor of the twist net machine, moved his entire business from Loughborough to Tiverton in Devon, where his large factory and its associated housing can still be seen. At Chard in Somerset, there are other ornate factories opened by East Midlands lace manufacturers. Eventually, lace machines were able to produce fully patterned lace and the hand lacemakers, who had supplemented meagre agricultural wages through their industry, struggled to survive by diversifying their products.

The finishing of lace was as important as that of linen, needing considerable bleaching to satisfy the market. A large number of bleachworks had been set up near Nottingham for the cotton industry while chemical bleachworks were established in the town during the nineteenth century. The most elaborate of these was Lambert's in Talbot Street which was completed in 1863 with an ornamental façade and tower. Dyeing and trimming of lace was carried out in separate premises by lace dressers, while mending and packaging took place in the warehouse, often in well-lit attics. Marketing was previously London-based but transferred to Nottingham in the 1850s, resulting in the creation of the Lace Market, a warehouse development comparable with 'Little Germany' in Bradford. The elaborate architectural style of the these warehouses, particularly that of Adams and Page in Stoney Street, contrasts sharply with stark utilitarian façades of the tenement lace factories. Another elegant example is the brick-built Gothic-style lace warehouse which fronts the Anglo Scotian lace mills in Beeston near Nottingham.

Dyeing and finishing

The rapidly expanding textile industry, in all its branches, was a considerable stimulus to the chemical industry because of its requirements for soap for washing and chemicals for bleaching. Although some fabrics were marketed in a white state, fashion demanded colours, particularly in cotton textiles and knitwear. This process was first carried out using vegetable-based dyes, made by crushing plants such as saffron and woad as well as imported materials such as logwood, which produced a black dye. A water-powered mill for chipping and crushing logwood, the Albert Mill at Keynsham near Bristol, retains some of its original machinery despite conversion to housing. Mordants had to be used to fix the dyes, of which the most important were copperas and alum: the former was derived from iron pyrites and the latter quarried from shale outcrops occurring, for example, along the Yorkshire coast. The National Trust have consolidated and interpreted the important alum shale works at Ravenscar where the remains of the crushing plant and stone-lined tanks for leaching out the alum may still be found. As in the bleaching process, chemicals superseded natural dyes in the mid-nineteenth century, with aniline dyes being derived from coal tar, a by-product of coal gas production. These processes required large quantities of hot water, and dyehouses were usually situated by rivers or canals and equipped with large storage tanks and extensive boiler plant. Some textile mills possessed their own dyehouses, others made use of specialist commission dyers. The louvred roofs of dyehouses, designed to disperse heat and fumes, were a characteristic feature of most textile towns.

The production of cotton fabrics rapidly overtook that of wool and a large proportion was both dyed and printed. This branch of the textile industry had previously been centred in the London area, and in 1805 the River Wandle was described as 'the hardest worked river for its size in the world', having ninety mills on it by 1831. The two famous printworks of William Morris and Arthur Liberty were established on the river in Merton, and produced high-quality fabrics by block printing and engraved copper plates.[23] But the main industry developed in the Manchester area, in tandem with the cotton industry, utilising the new method of continuous cylinder printing by engraved copper rollers in the last decades of the eighteenth century. Specialist textile printers were established and a group of them were located in the Sett Valley south of Manchester, taking advantage of its ample water supply both for powering machines and for the necessary washing processes.

The vast increase in the output of cloth had repercussions on the manufacture of clothing. This had previously been either a domestic process or carried out by professional tailors and seamstresses, often working in cramped, backstreet premises. In the mid-nineteenth century, the increase in full-time employment for both sexes and a rise in purchasing power led to a demand for ready-made clothing. The invention of the band knife for cutting out several layers of cloth at once to a template, combined with the sewing machine, helped to meet this demand and large clothing factories were established. Shirts were manufactured in the linen-producing districts, while knitwear became a specialist branch of the hosiery industry both in Scotland and the East Midlands. Leeds, home of the band knife, became the clothing capital of England and its factories supplied chains of retail shops which became household names at the beginning of the twentieth century, such as Hepworths and Burtons. Clothing factories were established in other towns in the second half of the nineteenth century, making use of small steam engines to drive large numbers of sewing machines. Good examples of these factories can still be found in Stroud, while the Kettering firm of Wallis and Linnell erected a fine pedimented stone building in 1873 in the nearby village of Brigstock, now converted into offices.

Hats were an essential item of clothing until the middle of the twentieth century. Summer versions were supplied by the straw plaiters of the south-east Midlands, where the hat industry was centred on Dunstable and Luton. By 1851 there were 22,000 straw plait workers in this area, and children were sent to plait schools as soon as they could walk. They learnt straw plaiting but little else in crowded cottage workshops. Plait was sold either to dealers or in markets in the towns, and plait halls, similar to the Yorkshire cloth halls, were built in Luton, Hitchin and Dunstable in the second half of the nineteenth century. Hats were made up and finished either in garden workshops or urban factories.[24] The manufacture of felt hats was centred near Manchester, particularly in Stockport and in the town of Denton, where in 1900 half the population were employed in the industry. Like textile weaving, felt hat-making was combined with agriculture until the industry became factory-based in the mid-nineteenth century. In the Stockport area, several large manufacturing complexes developed in the later decades of the nineteenth century, among them Christy's and Battersby's who supplied bowlers, trilbys and homburgs to the London market. There was also an important outpost at Atherstone in Warwickshire, where several hat factories survive.

The leather industries

The production of leather has always been one of the most widespread of all craft activities, nearly every town possessing its own tannery and several shoemakers and cordwainers. Leather was used for footwear, clothing, upholstery and wall covering, saddles and harness as well as parchment. Once powered industry developed, leather found new uses for drive belts and moving parts of machines such as power looms. From being a craft industry located in the countryside or on the outskirts of towns because of the noxious fumes created by the tanning process, leather production became an increasingly specialised urban industry. Hide and skin markets were established alongside many cattle markets, such as that built by Nottingham Corporation at its East Croft depot. Leeds became a major tanning centre while large tanneries near ports, such as those at Bermondsey in London, were set up to process imported hides.

The heavy hides required for boot and harness manufacture were tanned after preliminary cleansing by repeated immersion in successive open pits containing increasingly strong liquor derived from water and oak bark, a process which could take as long as fifteen months. The bark was shredded in water-powered mills, located either near the source of the oak bark or alongside the tannery: an example can be seen attached to the tannery moved from Rhayader to the Welsh Folk Museum at St Fagans. The pits were normally contained in open-sided sheds in order to prevent dilution of the tanning liquors by rain but still to allow the fumes to escape. The tanned hide was dried slowly in rooms ventilated by louvres, which are the most characteristic feature of a tannery building. Good examples of tanneries may be seen at Downton in Wiltshire, alongside the River Avon, and Grampound in Cornwall, the latter being one of the last tanneries to use oak bark in the traditional manner. The lighter leathers, like sheep and goat skins, were tanned with a mixture of alum and oil so as to prevent discolouration. In the course of the nineteenth century, tanning in power-driven drums replaced the use of open pits and greatly speeded up the process, particularly when the use of new chemical tanning agents such as chrome were introduced. These machines, together with others for splitting and rolling the leather, were then housed in conventional factory buildings and the open tan pits gradually disappeared.

Some of the leather was dispatched to curriers and dressers where dyeing and finishing processes were carried out. The premises of both tanners and leather dressers were usually situated alongside a source of water. Two extensive dressing works remain in Nottingham, one alongside the River Trent and the other by the River Leen, although both have now been converted to residential use. Some of the soft finished leather was used for glove manufacture, which became a specialist industry in parts of the south-west. One centre was the Somerset town of Yeovil, with many small factories being established in the surrounding villages. Walsall in the West Midlands was the country's most important centre for the manufacture of saddlery and horse harness, manufacturers being attracted there by the local production of lorinery or saddler's ironmongery. The Walsall leather industry was carried on, as was typical in the West Midlands, in a mass of tiny and often primitive backyard workshops, intermingled with a handful of larger factories. One of these, built in 1891, now houses the Walsall Leather Centre and Museum. The disappearance of the horse from our roads encouraged the town to diversify into the production of other leather goods such as the manufacture of handbags and other luggage, something which is still carried on.

The dominance of Northamptonshire in the manufacture of boots and shoes is not easy to account for, but as early as the Civil War it was supplying government orders for army boots. The county continued to specialise in the production of men's boots and shoes but, due to its reluctance to accept mechanisation in the mid-nineteenth century, the manufacture of ladies' and children's footwear by machine became established in neighbouring Leicestershire. Even so, much of the footwear industry remained home-based even longer than the hosiery industry. Production was split into separate processes and partly finished goods were transferred from one place to another in what became known as the basketwork system. Cutting out the leather, or clicking, was a highly skilled process and the first to be removed from the domestic workshop into warehouses, multi-storey buildings with wall-mounted cranes for lifting the bales of leather. Closing, or assembling the upper components together, was generally carried out in backyard workshops included in the rent of the house, like the topshops of the hosiery industry. These were single storey and often built as lean-tos against the garden wall, as can still be seen in the areas of Kettering, Rushden and Rothwell which were developed especially for housing boot- and shoe-workers in the last three decades of the nineteenth century. Hand sewing machines could be used in the domestic situation, but it was the introduction of powered sole-stitching machinery for lasting, or attaching the sole to the upper, which finally brought domestic workers into the factory environment. The typical footwear factory of the late nineteenth century was of three storeys and a basement, where leather was stored in dark and cool conditions. Clicking and closing took place on the top floor where there was maximum light, with lasting on the second floor and packaging and dispatch on the ground floor. As shoe machinery became heavier, firms were obliged to remove some processes into single-storey units as in the textile industries. Some factories, such as the Excelsior Works in Irthlingborough, demonstrate this transition, with a three-storey factory fronting a range of single-storey buildings to the rear.

Only a few boot and shoe firms expanded into vast premises: in Northampton these included Barratt's 'Footshape' Works, a magnificent edifice with buff terracotta ornamentation (Plate 21), and Manfield's extensive works of 1892 on the Wellingborough Road, in which all processes were accommodated on one level behind a 400-feet street frontage. In Leicester, it was claimed that the CWS Wheatsheaf Works were the largest shoeworks in the world when they were opened in 1891, ultimately employing 3,000 people.[25] Wholesale warehouses were also built, some to service their own chains of retail shops, and others of more elaborate style to impress buyers. Several of the central streets in both Leicester and Northampton were lined with warehouses, the ground floors of many of them having been converted to retail shops.

Like textiles and hosiery, the footwear industry brought ancillary trades in its wake. Packing was required, both in the form of baskets or skeps for use in the industry, and boxes for packaging the finished goods. Basket-making became very important in the valleys of the River Trent and its tributaries well into the twentieth century, using locally-grown osiers and even surpassing the production of the Somerset Levels. Waste leather was ground and used for the manufacture of leatherboard, several cornmills on the River Soar in Leicestershire switching to its production. The fashion for elastic-sided boots in the mid-nineteenth century created another industry, the production of elastic web. This was a knitted fabric with a cotton warp and

Plate 21 The elaborate Footshape Boot Works in Northampton which fronted north-lit single-storey workshops at the rear. On this site all the mechanised processes of shoe manufacture were gathered together. The products were marketed by a mail order system and also through the Barratt retail shop chain.

rubberised weft which was also used for the corsets fashionable in the Victorian period. The attractive Bridge Mills in Quorn, in Leicestershire, were built for elastic web and have round headed cast iron windows and a chimney which is still a landmark in the Soar Valley. In Leicester, St George's Mills in Humberstone Road produced elastic web as well as shoe components or mercery, which were marketed from an elaborate warehouse in nearby Rutland Street which is faced with marble and terracotta. Corset manufacture was important on the Leicestershire–Northamptonshire border, particularly in the elaborately fenestrated factories belonging to Symingtons in Market Harborough, whose products are now on display in the Snibston Discovery Park at Coalville. The Co-operative movement was very strong in Northamptonshire and the CWS operated a large corset factory in Desborough.

The clothing industry had, perhaps, a more far-reaching effect both on the lifestyle of those working in it and on the built environment than any other industry. In the course of the two centuries we are considering, the manufacture of textiles and clothing ceased to be a craft practised in scattered settlements, combined with farming, and became one carried out almost entirely in a factory environment. But study of the landscape reveals that this was not a process which followed the same pattern in every industry nor in every region of the country. What is perhaps surprising is how long the process was delayed in, for example, wool and cotton weaving, hosiery and the boot and shoe industry. Our impression of the 'factory revolution' in textiles has been created by contemporary reports and engravings of the novel structure of Arkwright's factory at Cromford or the conditions of the mill-workers in the Lancashire cotton towns. What these sources do not indicate, but the study of maps and buildings does, is the length of time that weaving, knitting and stitching leather continued to be practised in a domestic environment or in purpose-built workshops which were being constructed until the very last decades of the nineteenth century. Landscape evidence does, in fact, modify very considerably our view of the wholesale and rapid mechanisation of the textile industries.

The landscape is also very revealing of the different methods of industrial organisation in the clothing industries and the changes they underwent during the two centuries we are considering. The urban warehouses and village-based weaving and knitting workshops, together with networks of packhorse tracks and tiny lanes, illuminate the nature of the putting-out system in the woollen and hosiery industries. Even this, of course, varied regionally, the laithe houses of the independent Yorkshire yeomen clothiers contrasting with the elegant town houses of the capitalist clothiers of south-west England. Thomas Lombe's Derby silk mill was a premature signal of a change in organisation which was to affect several sectors of the textile industries before the end of the eighteenth century: the use of a central power source around which both the processes and the workforce needed to be grouped. The use of water-power, often in isolated areas on the slopes of the Pennines, created the model communities which appear to have characterised the cotton spinning industry in the last two decades of the eighteenth century. But these were not typical of the industry as a whole, and the advent of steam power brought the textile industry back to the urban environment in which it had flourished in the late Middle Ages. Living conditions in nineteenth-century towns were often worse than those experienced by the workforce during the water-powered phase of the industry.

The buildings themselves reveal much about the technological and industrial organisation of various branches of the industry. A general transition can be seen from manufacture carried out in the home to the provision of separate but still hand-powered workshops and finally the factory with its central power source. The buildings also indicate the increasing complexity and weight of the machinery which came into use in the nineteenth century, with weaving looms and sole-stitching machines needing to be placed in single-storey premises. Multi-storeyed mills increased in floor area to accommodate larger mules and ring spinning machines, but this was only possible through improved building techniques. Fire, of course, was a particular hazard in the textile and leather industries, and it was textile manufacturers who pioneered the use of large, fireproof structures incorporating new materials such as cast iron and reinforced concrete. These were then utilised by other industries, notably transport and food-processing.

The clothing industries, too, provided a stimulus to other industrial activities, including iron for the manufacture of machinery, coal for steam power, chemicals for bleaching and dyeing, baskets and boxes for packaging and printing for the identification of goods. Textile and leather towns developed a range of ancillary activities of this kind, together with engineering derived from the skills of building and maintaining machinery. The manufacture of bicycles in Nottingham, cars in Coventry and rubber in Leicester were all spin-offs from the textile trades and enabled these towns to survive the cyclical slumps which bedevilled the clothing industries. Finally, the large-scale production of cheap cloth and the ready-made clothing industry closed the social gap in the mid-nineteenth century. The range and quality of the wardrobe possessed by a mill-owner's wife obviously far surpassed that of his workforce but even the millgirls now had their Sunday-best clothes. By the end of the century, the textile and leather industries were certainly 'clothing the people'.

6

Building and servicing the community

In the period under consideration, more land than ever before vanished beneath a blanket of building: new settlements were created, existing ones enlarged and vast urban conurbations appeared in hitherto sparsely populated areas, bringing with them a host of new problems concerned with public health and hygiene. This expansion of building was caused partly by the unprecedented growth of population and partly by its redistribution to meet the needs of industry. The comparatively small settlements of the seventeenth century were still characterised by buildings in local vernacular styles using easily available materials. Once transport by canal and rail enabled bulky products like bricks and Welsh slate to be competitive, the regionally distinct styles of building were gradually replaced by the monotonous uniformity which marked so much of the nineteenth-century urban landscape. Only close inspection reveals the quality of much urban housing of this period and the decorative detail with which it was sometimes enhanced. The nature of the townscape was also influenced by political consideration. While governments had previously assisted towns ravaged by fire, they now saw the building boom as a source of revenue and taxed different classes of building materials, thereby limiting the choice available to those unable to meet the additional cost. Only with the public health problems of the mid-nineteenth-century towns did both national and local governments adopt a positive attitude by passing legislation to regulate the development of the built environment. This chapter will consider, firstly, the provision of building materials, secondly, the changing styles of building in the period and, thirdly, the social problems created by urbanisation and their solutions in the form of legislation and the provision of public services.

Stone and the slate

The pre-nineteenth-century buildings of Britain owe much of their distinctiveness to the immense variety of available building stone within such a compact area. This is revealed by a close study of many public buildings, the materials used indicating the extent of the transport network which provided them. The London Guildhall, for example, has Kentish Rag walling and Collyweston slate roofing; the Royal Exchange of 1838 utilises Cornish granite and Portland stone. Igneous, metamorphic and sedimentary rock were all used for construction purposes, and quarries, both large and small, were features of the stone regions which were mainly in the north and west of the country. During most of the period, quarrying relied mainly upon manual labour, although gunpowder was used for blasting and both powder stores and quarrymen's blast shelters can still be found in abandoned quarries. The stone was lifted with the assistance of wooden shearlegs and cranes, but dressing remained largely a manual process until the introduction of mechanical saw frames. Horses were used for haulage to the nearest canal or seaport, and quarries within easy reach of the coast had a competitive advantage.

Granite, the most intractable of igneous rocks, was locally used in the random state for building, like the pink Mountsorrel granite which can be seen in the villages of the Soar Valley in Leicestershire. In its dressed form, granite was popular for buildings requiring strength, such as the Cornish engine houses, where both the quoins and cylinder bedstones made use of local granite. Its use was more widespread for civil engineering purposes, such as bridges and viaducts, docks and harbours. The Trustees of Liverpool Docks leased a 30-acre quarry at Kirkmabreck, Kirkcudbrightshire, in 1830 to provide materials for Jesse Hartley to construct their new docks.[1] There they built a loading jetty, cranes and railroads to ship out the stone. The streets of the burgeoning towns were also paved with granite setts and kerbs, particularly after canal and rail transport made this economic.

The quarries which scar Dartmoor and elsewhere in south-west England provided granite for buildings and civil engineering structures, especially in London, to which it could be sent by sea. Many of London's bridges over the Thames were built of granite from Devon and Cornwall, and the de Lank quarry provided John Smeaton with the materials for the new Eddystone Lighthouse in 1759. The grim edifice of Princetown prison was erected from local granite by French prisoners during the Napoleonic Wars and the nearby Merrivale quarry continues to provide building stone.[2] One series of quarries at Haytor on Dartmoor is notable for its tramway system which carried granite to the Stover Canal and the River Teign. Built as late as 1820, the system is unique in that instead of the more usual iron rails, the Haytor Tramway is constructed with carefully shaped 'L' rails and points hewn from granite blocks (Plate 22). The stone was used for the rebuilding of London Bridge in 1825 and for work on the British Museum and National Gallery.[3]

Slate, the principal metamorphic rock, found more widespread use for roofing than building, except in the immediate vicinity of the quarries where slate-built barracks and housing blend into the landscape. Its universal use on terraced housing in nineteenth-century townscapes was due to a combination of factors, mainly the availability of rail transport, coastal shipping and the lightness of the slates which enabled economies in roof timbering. The output of the slate quarries of Caernarvonshire, for example, increased from under 20,000 tons in 1786 to over 90,000 tons by 1831.[4] Several parts of North Wales are still totally dominated by their nineteenth-century past: galleries or benches are piled on top of one another high up the mountainside and the waste spills over into the valleys below. Blaenau Ffestiniog is flanked on three sides by old quarries and by mountains hollowed out in the quest for slate, with rail inclines topped by towers supporting brake drums. Rhosydd Quarry still preserves the remains of its slate-dressing mill, powered by water stored in reservoirs high above, together with smithies, offices and further rows of barracks.[5] The landscape

Plate 22 A section of the Haytor Tramway which was built on Dartmoor to move granite from the Haytor quarries, seen on the skyline, to the Stover Canal for shipment. This primitive railway was built in 1820 and used granite blocks up to 8 feet in length with rebates on their outer edges instead of rails. Horse-drawn waggons with flangeless iron wheels operated for forty-five years. The quarries, which supplied stone for London Bridge and other London buildings, continued working until 1865.

is perhaps most easily interpreted at Dinorwic Quarry at Llanberis, the horizontal terraces rising 1,400 feet up the slopes of Elidir, only broken by the vertical stripes of inclined plane railways for lowering slate to the workshops which are now the North Wales Quarrying Museum. These were built in 1870, the machinery being driven by an overshot water-wheel of over 50 feet in diameter, now one of the largest surviving in Britain.

Slate was also quarried in the Lake District and at Delabole, in Cornwall, where it is still extracted from a vast excavation begun in the late Middle Ages and now 500 feet deep. The slates from here were carried by horse and cart to the small harbours of Port Isaac and Port Gaverne until the arrival of the North Cornwall Railway in 1893. Generally, slate was sawn and split in the quarries; it could be cleaved very thinly and dressed to standard sizes, hence its popularity for the large-scale development of towns. Local slates held their own by supplying a different market, the thin-bedded sandstones of the Pennines and the limestone of Collyweston in Northamptonshire and Stonesfield in Oxfordshire being used for roofing substantial farmhouses and other domestic buildings. In these quarries, blocks of stone were left to weather over the winter, the frost action making cleaving easier when spring arrived.

Sedimentary rocks, which include sandstone, gritstone and limestone, are more widely distributed in the south and east of Britain and have been extensively used both for domestic and public buildings. Some quarries provide freestone for building blocks or for shaping as decorative features, while others provide stone which can only be used in random form and has a more local distribution. Among the latter is carstone, which along with chalk or clunch provide the only stone suitable for building in East Anglia, hence its great wealth of timber-framed and later brick buildings. The rusty carstone can be seen, for example, in the buildings of Downham Market, particularly the station and maltings nearby. Sandstone, like carstone and clunch, weathers badly, but despite this has been extensively used as a building stone. It ranges in colour from the dark stone seen in the walls of the little Nether Alderley water-mill in Cheshire, now in the care of the National Trust, to the buff stone of the unadorned but massive Shaddon textile mill in Carlisle. The Craigleith quarries, near Edinburgh, supplied the stone for many of Edinburgh's public buildings and also for several in London. Gritstone from the Coal Measures may be seen in abundance in the mill towns of the Pennines; now blackened by smoke, the mills and warehouses blend into the landscape from which they were hewn. The Millstone Grit of the Pennines, as its name implies, was also used for stones for corn mills, edge runners for crushing minerals or apples for cider and grindstones for the metalworking trades. Small quarries are located along the 'edges' or scarps of the gritstone in Derbyshire and Yorkshire, and unfinished millstones in various stages of manufacture can be discerned. Hundreds of small millstones, probably grindstones, are stacked near the

Bole Hill Quarry to the north of Grindleford in Derbyshire; they lie alongside the trackbed of the railway opened to transport materials for the building of the Derwent dams in 1907.[6]

Limestone and lime-burning

Limestone has always been valued as a building material, both for its colour and the ease with which it can be dressed and carved. It is a well-bedded rock, jointed into natural blocks, which could be removed with crowbars and wedges. Excellent building stone was quarried at Clipsham and Ketton in Leicestershire and Ancaster in Lincolnshire, where the remains of the quarries can still be seen. Bath stone was widely used for public buildings in London and, because of the thickness of the overburden, was mined rather than quarried. The Combe Down quarries near Bath were the scene of a very early railway, built in 1730 by Ralph Allen to carry stone down to the River Avon. Portland stone from Dorset was extensively used in the nineteenth century for exterior decorative work on buildings. Some limestones take a particularly good polish and are known as 'marbles', for example, the Purbeck marbles, again of Dorset, and Dent in Cumbria. Here the rock was hauled on sledges to hand- or water-powered mills where sawing and polishing were carried out. At Ashford-in-the-Water, in the Peak District, an impure form of limestone was worked which became black when polished. It was mainly used for inlay work which sold well in the nearby fashionable health resorts of Buxton and Matlock.

A great deal of limestone has been quarried for purposes other than building, including use as a flux in iron smelting, road-making and burning in kilns to produce lime. Once a very widespread industry, utilising whatever inliers of limestone were available, lime was in considerable demand both for improving soils when wastes and commons were enclosed and for making mortars and cements. The earliest method of burning limestone was in a clamp, but the kiln was more efficient. Alternate layers of stone and fuel were placed in a bowl-shaped open vessel lined with firebrick and burnt over a period of days. At the base of the kiln were one or more arches which enabled air to enter the charge and burnt lime to be raked out. Limestone was burnt either at the quarry, or, because of the corrosive nature of burnt lime, transported to where it was needed and burnt there. For ease of loading, kilns were often built against banks either as isolated structures in the landscape or in the bottom of quarries. The single kilns, once very common, are now mostly ruinous and overgrown, although some have been preserved as landscape features, such as the kiln worked by the poet John Clare at Pickworth in Leicestershire and the Bishop's kiln north of Abbotsbury in Dorset. Lime was also quarried and burnt on landed estates, both for their own use and for sale, and many landowners made a considerable profit on their kilns. On the Calke Abbey estate in Derbyshire, for example, the Harpur-Crewe family owned a series of quarries, each of which supported several kilns. The burnt lime was both used on the estate and sold to farmers: it was also distributed more widely via the Ticknall tramway, constructed about 1802 to allow access to the Ashby Canal. The tramway also carried limestone to supply kilns erected beside the canal which were closer to the source of coal.[7]

Limekilns became a common feature beside canals and railways in the course of the nineteenth century. At Froghall in Staffordshire, an impressive bank of kilns was built at the terminus of the Caldon Canal. They were supplied with stone by tramways and inclines from the extensive quarries on Caldon Low. The hills at Dudley in the West Midlands are honeycombed with limestone workings, served by underground canals. Some of the limestone was burnt in the large range of kilns at the eastern entrance of the Dudley Canal tunnel, now part of the Black Country Museum site. In the Peak District of Derbyshire, where large-scale quarrying still continues, unusual rock-cut kilns were constructed in 1880 beside the Midland Railway at Millers Dale and have been conserved by the National Park. The majority of limekilns were gaunt, strictly functional structures, but others showed architectural pretension, such as the range with Gothic arches at Tenby in Pembrokeshire or the even more impressive range beside the Llanelly Railway at Llandybie, near Ammanford, designed in suitably ecclesiastical style by R.K. Penson, a local church architect. Coastal shipping brought limestone and coal to kilns constructed in small harbours, such as those at Solva and Porthclais in Pembrokeshire and Clovelly in Devon. In north-east England, the National Trust have conserved sets of kilns at Seahouses, Beadnell and Lindisfarne. Towards the end of the nineteenth century, the continuous Hoffman kiln already in use for brick-making was utilised for lime-burning to supply mortar to the growing towns. Few of these survive, but one has been protected by listing at Langcliffe, near Settle in North Yorkshire.[8] Several other types of kiln built to foreign patents were introduced around 1900, particularly for burning chalk in south-east England and spectacular examples survive at Betchworth in Surrey and in the Chalk Pits Museum at Amberley in West Sussex. Most lime-burning is now carried out in steel rotary kilns.

The quarrying of gypsum, too, has undergone dramatic development since its origins in the late Middle Ages. It is a hydrated sulphate of calcium, usually soft but occasionally hard and veined with red or green: it was then quarried and polished as alabaster. Nodules of alabaster were worked underground from levels or shafts at Chellaston and Aston, near Derby, being widely used for effigies and church decoration. The softer gypsum was worked for plaster, at first in pits as at Tutbury in Staffordshire but then by underground working. Plaster was used both for covering walls and floors and for decorative purposes, such as the pargetting which is common on the timber-framed houses of East Anglia. By the end of the nineteenth century, large gypsum quarries were opened, for example, near Newark in Nottinghamshire, and finally continuous beds of the dehydrated form of gypsum, anhydrite, were discovered in the Vale of Eden and Teesside. These have been quarried for plaster, cement and as a raw material for the chemical industry.

A more substantial flooring material was concrete, a mixture of sand, cement and water; the cement was made by burning a mixture of limestone and clay at a higher temperature than was used in a limekiln. Cement kilns were therefore bottle-shaped like pottery kilns to achieve these temperatures, and more than 1,000 could be found around 1900 in the Thames and Medway areas where suitable clayey limestone occurred. Concrete was first used for foundations for both buildings and machines but reinforced concrete construction utilising wrought iron bars was possible from the middle of the nineteenth century. William Fairbairn pioneered its use for textile mill construction, but the conservatism of many mill-owners prevented its widespread adoption and brick remained the more usual material. Developments in Europe towards the end of the century, such as the Hennébique system, perfected reinforced concrete as an acceptable building medium. It became widely used for large structures such as grain silos and warehouses, one of the first applications of the Hennebique system being the grain warehouse built for Weaver & Co in Swansea Docks in 1898. Concrete became an almost universal replacement for iron, brick and stone in civil engineering works but does not blend into its surroundings as well as natural materials. Concrete accretions on mine and quarry sites are alien elements in the landscape, intruding into previously harmonious settings where only indigenous materials were used.

Bricks and tiles

Many areas of Britain had no resources of good building stone and relied on imported freestone for churches and public buildings, domestic buildings generally being timber-framed. The depletion of timber reserves from Elizabethan times prompted the search for alternative materials in these areas, particularly flint and clay. The latter was used in cut-block form or mixed with straw and water and allowed to set between wooden shuttering, forming the 'cob' so common in Devon cottages. However, it was burnt clay, in the form of bricks, which became the main building material in these areas and came to dominate the British landscape by the latter half of the nineteenth century, except where stone prevailed. The Romans had made bricks and tiles in Britain, but their use declined thereafter until the thirteenth century when bricks were imported from the Low Countries and used in prestige buildings. Local brick manufacture began where suitable clays and fuel could be found, brick replacing the wattle and daub filling in timber-frame construction and finding a use for the chimney stacks being inserted into open hall houses. Brick building spread downwards through the social scale from the manor houses of the fifteenth and sixteenth centuries to the labourers' cottages of the nineteenth. Fashionable brick façades disguised timber frames in urban settings, while legislation following the Great Fire of London in 1666 attempted to enforce the use of brick and stone for wall construction in the capital. Bridges, locks and warehouses of brick dominated the canal scene and a rash of brick building spread across the country once canal and railway transport enabled coal supplies to be brought to areas with suitable clays. It is scarcely surprising, therefore, that the manufacture of bricks and tiles became a prime target in government attempts to raise taxation at the end of the eighteenth century. A tax on bricks of 2s.6d. per 1,000 was first introduced in 1784, prompting some entrepreneurs, notably Joseph Wilkes of Measham in Leicestershire, to produce double-sized bricks, an evasion countered by the government in 1803. Mathematical tile hanging and weatherboarding were alternative methods of avoiding payment. The tax was finally removed in 1850, by which time the canal and railway companies had made a major contribution to government coffers.

The most common use for clay was in the manufacture of bricks, but curved pantiles for roofing were introduced into East Anglia in the seventeenth century as a replacement for thatch. Patterned floor tiles had been a feature of medieval monastic churches, but were copied in the nineteenth century for a wide range of domestic purposes, particularly by firms in the Potteries and by Maws and Craven Dunnill at Jackfield near Ironbridge. Other forms of decorative clay work were rubbed and moulded bricks, often used for the cornices and lintels of terraced houses, and terracotta, a fine-grained clay mixture which could be moulded into intricate patterns to replicate carved stone. This was either used unglazed after a single firing or suitable glazes applied before a second firing to produce faience. Terracotta was valued for decorative purposes on public buildings, although dated panels of terracotta work and ornamental chimney pots occasionally enhanced artisan housing. Fireclays, found in the Coal Measures, were made into bricks for furnace linings, as at Stourbridge whose products supplied the Black Country iron furnaces. Other types of clay, particularly that found in the South Derbyshire and Leicestershire coalfield, proved suitable for the manufacture of sanitary ware and salt-glazed drainpipes. The area produced most of the drainage pipes made necessary by the public health reforms of the mid-nineteenth century.

The almost infinite range of colours of bricks in the built environment reflects both the great variety of clays and the different firing methods employed. Red tones are perhaps most common, as, for example, in the stark, small windowed Rykneld Mills in Derby and other silk mills in Leek and Macclesfield. Bright red bricks predominate in the textile towns of Lancashire, many produced by the Accrington Brick and Tile Company.

Their 'NORI' brick eliminated the need for pointing, the frog or indentation in the brick being filled with lime mortar and the bricks pressed together in the building process, resulting in a smooth-faced wall. The clays of chalk and limestone areas produce a yellow, grey or buff brick, seen in large quantities in London and the Home Counties. In the late nineteenth century, the Lower Oxford clays near Peterborough began to be exploited on a large scale: they contained carbonaceous matter which reduced the amount of coal required for firing. These 'Fletton' bricks were widely distributed by rail and undermined the vernacular building tradition of many areas of Britain. The blue bricks used for engineering structures such as viaducts were made from clay with a high iron content and the firing method produced a dense, waterproof brick. These were also used in conjunction with other coloured bricks to produce the patterns which relieve the bland red brick of Victorian towns. A particularly fine example of polychrome brickwork is the Templeton's carpet factory in Glasgow, built in 1889 to resemble the Doge's Palace in Venice with red, light green and cream bricks (Plate 23). The range and quality of brick building in this period indicates the extent of the brick industry, but the actual evidence of production is more difficult to find.

Plate 23 One of the most striking textile factories is the former Templeton's carpet factory overlooking Glasgow Green. Part of a large complex, this Venetian Gothic four-storey plus attic works was built in 1889 using ornate polychromatic brickwork (Photograph reproduced by courtesy of Royal Commission on the Ancient and Historical Monuments of Scotland).

There were brick pits in nearly every parish in lowland England, but these can sometimes now only be traced through map and place-name evidence. Many are water-filled and used for recreational purposes in a similar manner to redundant sand and gravel workings. The clay was dug and allowed to weather before grinding in pug mills, often horse-powered like the crushing rollers which survive as part of an estate brickworks at Calke Abbey in Derbyshire. Bricks and tiles were made first by hand moulding but machines were introduced in the course of the nineteenth century, resulting in a more consistent product. The green bricks and tiles were then dried by stacking in long open-sided sheds before firing: these occupied a considerable area within a works and their pantiled roofs are a characteristic feature of surviving small works such as those dotted along the south bank of the Humber. Many works employed artificial heat to accelerate the drying process, but this was unnecessary for the Oxford clays as the bricks could be fired immediately after moulding without any previous drying.

The attractive random colouration of early brick buildings is a result of uneven burning in the earliest form of kiln, known as a clamp. This was not a permanent structure, the dried bricks being stacked on a clay base and covered with turves: the whole heap was fired with brushwood or coal and then allowed to cool. The clamp continued to be used into the nineteenth century where large quantities of bricks were required, as, for example, in canal and railway construction where the clamps were constructed on site. The first permanent brick kiln was probably the Scotch kiln, an open-topped rectangular chamber in which the bricks were stacked to leave passages for hot air circulation through them: the kiln was fired through a series of apertures down each side and the bricks covered with turves during firing. These were often used on country estates and may occasionally still be found in a derelict state, as at Calke Abbey where the kiln continued in use until closed by black-out regulations during the Second World War. The Suffolk kiln was similar but often built into sloping ground with fire tunnels below the floor of the kiln. A much repaired and modified example is still in use at the South Cove brickworks in Suffolk. Closed kilns presented problems in loading and unloading but were more economical in use. The most common type was the circular 'beehive' down-draught kiln, which was fired around the circumference and the hot air drawn through the kiln to a chimney. Single kilns served small communities, such as the well-preserved example at Bailey's Hard in Hampshire. Groups of these kilns were often connected by underground flues to a single chimney stack and their presence is easily detected from large-scale map evidence. Many still survive, having been converted from coal- to gas- or oil-firing. In the north-east of England, a rectangular closed kiln was used for firing both common and firebricks: this Newcastle kiln used horizontal draught and was capable of reaching high temperatures. All these kilns were intermittent in operation, requiring to be loaded, fired and then cooled, a process taking several days.

With the increasing demand for bricks in the mid-nineteenth century, continuous kilns were introduced in which waste heat from cooling bricks was used to steam green bricks before firing. This was achieved by means of a series of chambers arranged around a central flue so that the firing process proceeded sequentially from chamber to chamber. The continuous kiln was first patented by Hoffmann in 1858 and built in either a circular or elliptical plan. Large production units such as these were only constructed in yards with extensive clay reserves, and beehive kilns remained in use in small brickyards. Few Hoffman kilns are still operational, having been replaced from 1900 by the large transverse-arch or Staffordshire kilns and later by tunnel kilns, both of which today dominate the Oxford clay and other brick-producing areas. A Hoffman kiln continues in use at Normanton in West Yorkshire, while a disused kiln remains on the mining museum site at Prestongrange in Scotland.

Pottery and glass

Clay was not only used for providing building materials but also as a constituent of pottery. The medieval industry had utilised local clays of many different colours, but in the course of the eighteenth century demand for white pottery resembling imported porcelain increased. The ball clays of Dorset and Devon, quarried around Purbeck and Bovey Tracey, helped to produce a white body, but it was 'china clay' that satisfied mass demand. This type of clay, known as kaolin, is derived from decaying granite and extensive reserves have been exploited in the St Austell area of Cornwall. Pits were excavated and the soft clay washed down the face before pumping to settling tanks on the surface. The clay was then dried in specially-built pan kilns, with an underfloor heating system like a Roman hypocaust. Steam power was used for pumping the clay slurry, and some engine houses survive, notably that at Parkandillack which is dwarfed by the waste tips around it (Plate 24). These consist of waste quartz and mica separated from the clay and are the most characteristic feature of the St Austell area, often known as the Cornish Alps. Because of the lack of local coal supplies, the china clay was transported by a system of railways, coastal shipping and canals to the well-established Potteries in Staffordshire.[9]

Other ingredients were also used to whiten the clay body, notably flint and bone. The former was transported from the south coast and East Anglia to the Potteries by sea and canal, calcined in small kilns and then ground in water-powered edge runner mills. The remains of these can be seen along the River Churnet in Staffordshire, but the best surviving example is the Cheddleton Mill beside the Caldon Canal, dating from the 1780s. There are two separate mills on either side of the River Churnet, each with a low breastshot wheel driving the grinding pans. Bone was similarly calcined and ground, and the process can still be seen at the Etruscan bone

Plate 24 The landscape of the Cornish china clay industry with the Parkandillack pumping engine in the foreground. Jets of water were used in the clay pits to wash out the china clay or kaolin as a slurry which was then pumped to surface where the heavier sand and mica wastes were separated out in settling pits. The kaolin was then dried and packed for shipment, while the waste was deposited on conical tips which produce this classic Cornish scene. The Cornish pumping engine at Parkandillack remains *in situ*.

mill built in 1857 alongside the Trent and Mersey Canal near Stoke-on-Trent. Power for the grinding pans is provided by a small beam engine housed in a Dutch-gabled engine house at one end of the building. At the other, a massive square stack provides draught for two calcining kilns.

Beehive kilns had been used for firing pottery since the Middle Ages, but the distinctive pottery kiln or bottle oven was developed during the eighteenth century to provide the conditions necessary for firing glazed ware. Heat from furnaces passed through the inner structure or oven in which the pottery was placed, the draught being produced by the funnel effect of the bottle-shaped outer cover or hovel. Glazed pottery was normally placed in clay containers or saggars during firing. Some bottle ovens seem to have produced bricks and tiles as well as pottery, such as the preserved examples at Nettlebed in Oxfordshire, Luckington on the Badminton estate in Wiltshire and Bardon Mill in Northumberland. This type of kiln was also used to fire ceramic tiles, as in the large ornamental tileworks of Craven Dunnill and Maws in the Ironbridge Gorge (Figure 7). By far the largest concentration of bottle ovens was in the Potteries district of Staffordshire, where they were grouped in potbanks and surrounded by other buildings for clay storage and preparation, decoration and packaging.[10] Many of these were multi-storeyed with elaborate street frontages, as can be seen along the Uttoxeter Road in Longton, although the bottle kilns behind them have been demolished or replaced by modern tunnel kilns. At the Gladstone Pottery Museum, however, the bottle ovens do remain, but modern pollution regulations prevent their being fired. Traditional pottery techniques are still demonstrated and the atmosphere of a small, nineteenth-century potbank prevails. Many potworks were situated alongside canals, since water transport was ideal for the fragile product. Viscount Torrington, during his visit to Stoke-on-Trent in 1792, encapsulated the scene: 'these intersecting canals, with their passing boats, their bridges, the population, the pottery ovens, and the bustle of business, remind me of a Chinese picture, where the angler is momentarily interrupted by a boat'.[11] The surviving Coalport works, sandwiched between the River Severn and the Shropshire Canal near Ironbridge, presented a similar, although quieter, scene. The remaining kilns are now utilised to display the fine china once produced there. Other bottle kilns have been preserved at Bovey Tracey in Devon, the sole survivors of an important industry making use of local ball clays.

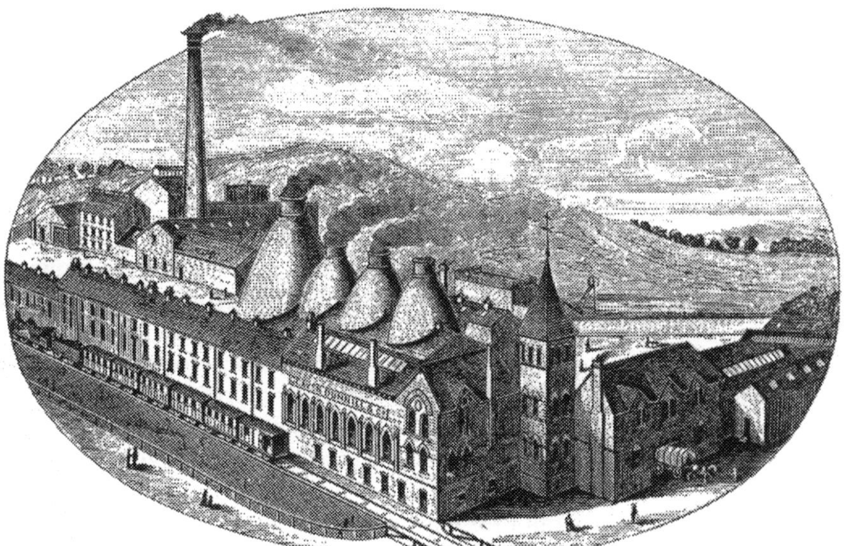

Figure 7 Craven Dunnill's tile works at Jackfield, near Coalport in Shropshire, around 1875. The Severn Valley Railway, opened in 1862, passes in front of the splendidly ornate works which made the encaustic and decorative tiles so much in demand during the Victorian era. Four coal-fired kilns can be seen, while heat and light was provided by coal gas from the works in the background. Now part of the Ironbridge Gorge Museum, the bases of the kilns have been excavated and tile manufacture is demonstrated to the public (Illustration by courtesy of the Ironbridge Gorge Museum Trust).

Another concentration of potworks for a more mundane market were situated on the South Derbyshire coalfield around Swadlincote and Church Gresley. Originally established to produce cheap domestic pottery in the 1790s, their output was transformed by the demands of public health reform in the 1840s. Salt-glazed drainpipes and sanitary ware were in great demand in towns and cities, and the fireclays of this area were ideal for the purpose. Numerous new potworks were opened, some with free-standing bottle kilns and others in which the kiln was partially enclosed in a multi-storey building. An extensive railway network was established to serve the industry, and numerous rows of cheap terraced houses were built on the very edges of the open clay and coal pits. The whole presented one of the most depressing industrial landscapes of Britain, although it is now in the process of reclamation.[12]

Glass-making kilns of the eighteenth and nineteenth centuries, known as cones, were very similar in both shape and function to pottery bottle ovens although considerably larger. They were essentially a truncated cone up to 80 feet in height, with numerous apertures around the base to create the necessary draught. This cone enclosed both the melting and annealing furnaces, while leaving space around them for the glass-workers to operate. Of the many which were constructed, only five survive in a recognisable form: the most accessible of these is at Stourbridge, once an important centre of the glass industry, and now forms part of a small museum. Another cone at Catcliffe near Sheffield was built in 1740 and has been retained as a landmark.

Inside the cone, a mixture of sand and a flux, such as soda ash, together with other additives, was fused at very high temperature in a clay crucible to produce glass. The molten glass was then shaped by blowing or moulding, but needed frequent reheating or annealing and so the glass-blower had to work inside the cone. Window glass was made either by blowing a glass bubble and spinning it into a disc which was then cut to form panes of crown glass, or swinging it to form a cylinder which could be cut and flattened. The slightly curved panes formed by these processes can be detected in the leaded lights and sash windows of old property. Like bricks, glass had been subjected to excise duty, which was removed in 1845 and the Window Tax was repealed in 1851. These changes enabled glass to be used extensively, not only for house and factory windows but for glazing the vast new iron canopies being built for market halls and railway stations. Its most spectacular use was in the Crystal Palace, built for the Great Exhibition of 1851 to the designs of Joseph Paxton, who had already created the great conservatory at Chatsworth House for the Duke of Devonshire. The glass for both of these structures was produced by the cylinder method by Chance Brothers of Smethwick. Already, however, large sheets of plate glass were being produced in the Ravenshead Works of the British Plate Glass Company at St Helens on Merseyside and this began to displace the older methods of making window glass. Plate glass, being highly polished, was suitable for mirrors and display windows for shops: the remarkable Arighi Bianchi furniture store in Macclesfield, built in the 1880s, is a good example of the latter.

Industrial communities

The most dramatic additions to the built environment in the period we are considering were factories and mills. Previously, the only structures on a similar scale had been churches, cathedrals or country mansions. Manufacture had generally been carried out in domestic premises or small-scale workshops. The factory concentrated most of the processes of production on a single site and for the first time brought a great many people together under one roof. Embryonic factories in the form of textile workshops had previously existed, but the use of a central power source became an essential characteristic of factory production. Factories were utilised not only in the textile industry, but also in iron manufacture, brewing, pottery and several other industries.

Towards the end of the eighteenth century, textile mills appeared in the most unlikely places, often isolated both from settlements and the primitive transport network then existing. A prime reason for this was the need for a good water-power site, with an adequate flow and a reasonable fall of water. Some of these sites were already in use as corn mills, fulling mills or forges and were converted to other purposes. In the corn mill, power was transmitted from the water-wheel via gearing and a vertical shaft to the stone floor, and this principle was extended to provide drive for the various floors in the textile mill. New mills sprang up in river valleys, but as the number of mechanised processes increased, either the water-power became inadequate or the full range of production could not be accommodated economically. Many previously water-powered mills had steam engines added to them before the end of the eighteenth century, some to pump water back over the wheel and others to supplement the water-wheel by driving machinery directly in times of drought or frost. However, once the rotative steam engine had been perfected, factories were better sited close to coalfields or to transport networks along which coal could be carried. The flexibility of the steam engine and its greater power output enabled new textile machinery to be developed, such as power looms and spinning mules: it also allowed the introduction of large rolling mills in ironworks, increased output in tower breweries and the mechanisation of many other industries such as pottery. The availability of coal to provide steam for power also enabled the factories to be heated, while the production of gas in factory plants meant they could be lighted too. The constraints of daylight working became a thing of the past and the workforce were compelled to accept long hours and shift-working systems.

The working conditions of many of the labouring classes, previously used to the domestic system, changed dramatically within their lifetimes.

The first textile mills sprang up in idyllic river valley settings, built in the vernacular style of locally-available materials. In appearance, they differed little from domestic buildings with large numbers of small, segmental headed windows. Their effect on the landscape was heartily disliked by aristocratic travellers such as Viscount Torrington, who felt that Sir Richard Arkwright's schemes had crept into every pastoral vale and destroyed the course and beauty of nature.[13] Few of these mills had any architectural pretensions, except where they expressed the social aspirations of their owners: Arkwright's Masson Mill has elaborate Venetian windows and William Newton's Cressbrook Mill, replacing an earlier Arkwright building, has an impressive classical pediment and bellcote. A striking example of this attitude is Castle Mill at Linby in Nottinghamshire where the façade looking towards Lord Byron's Newstead Abbey estate is castellated with quatrefoil windows, while the rear is totally plain. The earliest urban factories were also built in a utilitarian manner with regularly spaced small windows, but their appearance changed with the introduction of steel framing. No longer were the walls load-bearing and so the window area could increase, resulting in the large square glazed areas characteristic of late nineteenth-century factories. At the same time, the prosperity of the owners was reflected in architect-designed factories, often in the high Gothic style, although this was often confined to the street frontages. One other type of building which lent itself to this treatment was the tower brewery, and many examples survive of highly ornamental urban breweries such as the Castle Brewery at Newark-on-Trent with its château-style office block and red-brick brewing tower. The High Victorian factory no longer betrayed its industrial function, and could be multi-purpose: its line shafting could equally well drive a lathe, a knitting or sewing machine or a spinning mule.

Settlement patterns were also affected by changes in the location of industry. Some new settlements originated from the establishment of industries on rural sites where housing was non-existent, while others were begun by established industries whose very success had precluded expansion on their original site. Other settlements were related to the existence of raw materials or power sources: mineral wealth was there to be exploited or a suitable fall of water was available to provide power. A convenient transport system was also essential for bringing in or shipping out raw materials, whether it be river, canal, railway or access to a port. Canal undertakings themselves were to a great extent governed by topographical considerations: canal junctions, flights of locks and transfer warehouses often required living accommodation for their operatives. The same was true of railways, with staff to be provided at stations and junctions, while locomotive, carriage and waggon works of necessity needed to be adjacent to the tracks.

In order to attract labour to these new rural settlements, the employer had to provide a standard of housing normally higher than that enjoyed by agricultural labourers or town-dwellers. Many of the houses, particularly those associated with collieries and ironworks, had large gardens or allotments which cushioned the workforce against rising market prices and provided healthy outdoor activity in contrast to the confined working environment. The employer often also built social amenities in the form of schools, reading rooms and libraries and leisure facilities, together with places of worship. In Macclesfield in Cheshire, where a huge Sunday School was established as early as 1813, several of the churches and chapels were financed by local silk manufacturers, including the Anglican Christ Church which was paid for by Charles Roe in 1775–6.[14] The provision of beer houses depended on the attitudes of the entrepreneur, Quakers such as Joseph Rowntree and George Cadbury banning them from New Earswick and Bournville, whereas in Cromford, Richard Arkwright provided the splendid Greyhound Inn as a social centre. But these model communities had their less attractive side. Housing was often designed to reinforce the distinctions between different groups of workers, as can be seen clearly at Belper in Derbyshire where the foremen occupied groups of four attached houses in a single block, now known as the 'clusters', while the factory hands lived in the terraces of Short and Long Row. In colliery villages, the foremen often occupied a separate terrace named 'Deputy Row'. During the 1840s, the new railway town of Crewe was built with a considerable variety of houses, ranging from the villa-style lodges of the superintendents through detached houses for the engineers to closely packed terraces for the labourers. Occupation of a company cottage also allowed the employer to take a further hold on the tenant's life, preventing his taking any other job or even secondary employment. The employee and his family either worked for the company or found a new home. Many isolated textile communities also made use of pauper apprentice labour until prevented from doing so by legislation, after which the employer looked to the families of his workers for his piecers and bobbin winders.

The urban factory-owner was not obliged to provide decent housing to attract his workforce. The individual had greater freedom in his choice of employment but was largely at the mercy of greedy landlords or the speculative builder for his accommodation. Multiple occupation of houses was common, and many people lived in back-to-back or court dwellings and cellars. There were few gardens or allotments which could provide a subsistence living when there was no work to be had. Between 1800 and 1840, the quality of town life deteriorated as more and more people flocked into them in search of a job. The plight of the urban worker was highlighted in numerous reports and taken up by philanthropic institutions in large towns. In London, the Metropolitan Association for the Improvement of the Dwellings of the Industrious Classes was inaugurated in 1841 and in the next forty years provided over 800 dwellings and tenements as well as a lodging house for 300 men in East London. Many of the large drab tenement blocks erected by another philanthropic body, the Peabody Trust, still remain. Self-help as well as philanthropy was a feature of Victorian society and working men banded together in freehold land societies and building societies. The former were founded after the 1832 Reform Act to make it possible for working men to acquire a house and land so that they qualified for a vote in parliamentary elections. A number of permanent building societies grew out of freehold land societies, both encouraging saving and lending small sums of money to enable the working classes to build their own homes. However, private, speculative and philanthropic enterprise failed to provide the housing needed during the period of maximum increase in population and many townspeople continued to live in courts and cellars.

By the middle of the nineteenth century, however, public health problems led to changes in political attitudes and legislation on housing standards was enacted, if not always enforced. The Artisans' Dwellings Acts of 1868 and 1875 empowered local authorities to investigate and demolish insanitary housing, but there was no suggestion that the authorities themselves should rebuild houses; the cleared sites were handed over to private speculators. However, from 1858 local Public Health Boards were empowered to execute by-laws for new housing, but it was not until 1877 that model by-laws were produced recommending minimum standards.[15] These included regulations concerning street widths, depth of foundations, wall thicknesses, drainage and open space in front of new buildings and were generally adopted. The survival for over 100 years of many of the solid but uniform terraces of Victorian towns is largely due to the quality of design necessary to gain by-law approval. The hands of numerous builders may be detected in the different patterns used for lintels over doors and windows, the design changing between one block of houses and the next. It was not until the last decade of the period we are considering that local councils were empowered to construct working-class housing rather than leaving it to the private sector, but most did not do so until after the First World War.

The model rural industrial settlement was, in the last three decades of the nineteenth century, reproduced in an urban environment as the integrated industrial suburb began to spread outwards from existing towns. Streets of terraced houses, punctuated by corner shops, beer houses and multi-storey factories, were laid out in gridiron patterns, although the continuance of domestic industry was occasionally reflected in the rows of backyard workshops. Many of these suburbs were linked to the town centres by horse-drawn tramways or local railways and also enjoyed the provision of town gas. One example of such a suburb is North Evington, conceived by the architect–entrepreneur Arthur Wakerley on the outskirts of Leicester in the last two decades of the nineteenth century. New factories for the boot and shoe industry were built cheek-by-jowl with rows of terraced housing. Public facilities such as coffee shops, a fire station and a market hall were provided, but his efforts to create a self-contained community were frustrated by Leicester's new electric tramcar service which gave easier access to the town and encouraged the mobility of the labour force.

Municipal aspirations

The greater powers gained by the local authorities from the middle of the nineteenth century onwards encouraged new municipal aspirations, reflected in the townscape by the construction of town halls, assize courts and public baths. Town corporations were by this time strongly influenced by local entrepreneurs, whose spirit of private enterprise manifested itself in the architectural competitions held for the design of major public buildings. Many adopted Neo-classical styles, such as Hansom's town hall in Birmingham, dating from 1832, and the magnificent St George's Hall in Liverpool, designed by H.L. Elmes in 1839. In Todmorden in West Yorkshire, the wealth of the Fielden family of cotton spinners was reflected in the town hall, designed by John Gibson like a Roman temple and completed in 1875 on a scale far too large for the town. The Fieldens also built a Unitarian chapel and acquired several large mansions in the area, including the vast pile of Dobroyd Castle, constructed for them in the 1860s, again by John Gibson. Gothic styles became fashionable in the later part of the century, as illustrated by Alfred Waterhouse's designs for both the assize courts and the 1869 town hall in Manchester, while the town hall at Bradford, opened in 1873, boasted a magnificent Italianate tower. The commercial wealth of Cardiff was reflected in the exuberant Baroque of the city hall, part of the Cathays Park civic complex completed in 1901.

Some industrialists had previously set up schools in their factories and their interest in education was now shown by the foundation of mechanics' institutes, technical colleges and schools. Cultural life was equally well served by the provision of theatres, music halls, libraries and museums, many of which exhibited similar architectural grandeur to that of civic buildings. Health and welfare was provided for by public dispensaries, hospitals, asylums and workhouses, while the wealthy left a memorial of their opulence in the monumental masonry of private cemeteries such as Arno's Vale at Bristol, opened in 1840, and that at Undercliffe in Bradford, opened in 1851. Churches and chapels were built in the new industrial areas to serve the spiritual needs of the inhabitants, while their daily necessities could be purchased from shops, supplied from wholesale warehouses, and long-established markets which now often migrated into covered halls. Alongside industrial development went the commercial infrastructure of banks, corn exchanges, insurance companies and legal offices, all of which added new façades to the Victorian street.

The houses of the wealthy managerial and commercial classes also made their contribution to the townscape. Many had previously lived in close proximity to their work, like the West Country clothiers, but as their wealth increased they retreated to leafy suburbs, building large villas with coach houses and stables and living on equal terms with the professional classes. The expansion of the nineteenth-century towns brought about the depletion of many landed estates through the sale of land for both industrial and residential development, to the great profit of their owners. Some of them attempted to impose their will on new development through restrictive covenants, specifying either the size of building plots or the minimum building value of the houses erected. One example was Lord Calthorpe's estate of Edgbaston in Birmingham, developed from 1810 onwards to house the industrial and civic elite of the city. The restrictive covenants were not always successful in the long term; at Hillfields near Coventry the 10-yard frontage plots specified for a single dwelling were each eventually filled by three houses for silk weavers. In Leeds, the town elite moved out to elegant villas in Headingley and Roundhay, from where the city could be reached by horse buses. Probably one of the most coherent of these villa developments was the Park Estate in Nottingham, the former hunting park of the Dukes of Newcastle, where the local architects Hine and Fothergill gave free rein to their Gothic fantasies.

The spreading towns encroached upon the residences of the aristocracy who utilised the proceeds from the sale of land to escape from the smoke pollution of the factories to new country estates. In the West Midlands, the Fourth Earl of Dartmouth sold off most of the Sandwell Hall estate, which was 'much injured by smoke', to purchase a large estate at Patshull, 14 miles away from West Bromwich. The Wards of Dudley, whose fortune was derived from coal, iron and limestone, had settled at Himley Hall, south of the town, which by the 1830s was affected by local industrial development and regarded as unsuitable as a 'Residence for a Nobleman of considerable income'.[16] They added to their estates the magnificent Witley Court with its adjoining Baroque church 20 miles south-west of Dudley, where they entertained on a grand scale. By these means did those who had made their fortunes from the smoke seek to escape it.

Towns and public health

For the vast majority of the people, however, there was no such escape. Their numbers grew rapidly throughout the nineteenth century, the population increasing from 10.69 million in 1801 to 20.88 million in 1851 and to 37.09 million in 1901. The decadal rates of increase were between 11 and 17 per cent throughout the period, the highest periods of growth being between 1811 to 1831 and 1861 to 1881. In 1801 London was a major centre with one-tenth of the population, while four-fifths of the remainder lived in rural areas. By 1851, however, the population was roughly equally divided between rural Britain and those living in towns of more than 5,000 inhabitants. The proportion of town-dwellers continued to increase to the end of the century, but the growth of towns was by no means uniform. The population of Greater London increased nearly seven times between 1801 and 1901, while in the south-west the port of Bristol increased just over five times, while the nearby spa town of Bath only grew one and a half times. The increase was more than twentyfold in the woollen town of Bradford, and more than tenfold in Birmingham, Glasgow, Northampton and Southampton. Even more spectacular was the growth of entirely new towns such as Blackpool, which grew from almost nothing in 1801 to 47,000 by 1901, and Middlesbrough which grew from a similar base to 91,000 by 1901. In Manchester, the population trebled between 1774 and 1801, had nearly trebled again by 1831 and yet again by the end of the nineteenth century, largely due to the burgeoning cotton industry.[17]

For the most part townspeople were accommodated in poor-quality housing, often back-to-back cottages put up by speculative builders. These had party walls on three sides and were constructed in long terraces. Frequently, the spaces between terraces and back gardens were infilled to create the notorious court housing. In Nottingham, for example, in 1840 there were nearly 8,000 back-to-backs out of the total housing stock of 11,666, and the city possessed some of the worst court housing in Britain, even though these were given names such as 'Vine Court' and 'Pleasant Place'. Even in New Radford, a new industrial suburb of Nottingham where many two-roomed terraced houses were built, the average household size was 4.97 in 1841.[18] Overcrowding was common, several generations living in one room in a fetid atmosphere due to inadequate ventilation, water supply and waste disposal. The lack of cleanliness and personal hygiene also led to high mortality rates. Among other contributory factors identified by physicians in the eighteenth century were poor diet and the spread of disease through contagion. Some progress was made in the treatment of disease through segregation and scrupulous hygiene, together with the establishment of public dispensaries of medicines in the 1770s. The more risky innoculation against smallpox was replaced in 1796 when Edward Jenner introduced vaccination. Births and deaths were normally recorded in parishes, but it was not until after a major cholera epidemic in 1831–2 that the importance of medical statistics was recognised and the Statistical Society of London established to prepare information on public health, incomes and employment, housing, education and religion; only in 1837 was the office of Registrar General created.

Jeremy Bentham, the exponent of Utilitarianism, campaigned for public health legislation and through Edwin Chadwick, among other reformers, the means of improvement were seen to be better housing, provision for sewage and refuse disposal and a constant safe supply of water. Only in

1835 did the Municipal Corporations Act give some legal backing to the proposals, but inadequate finance and lack of agreement between adjacent local authorities delayed progress. The medical profession established the link between dirt and disease, but the solutions lay in social reform and the application of engineering. Chadwick, secretary of the Poor Law Commission, had in 1842 compiled his *Report on the Sanitary Conditions of the Labouring Population of Great Britain*, which with two other reports on England and Scotland prompted the establishment of a parliamentary select committee which recommended the enactment of the Building and Sewerage Acts with requirements that each town should have a Board of Health and a Public Health Inspector. A Royal Commission investigated the health of fifty large towns and its reports in 1844 and 1845 confirmed Chadwick's findings. The resultant Public Health Act of 1848 created a new executive branch of government in the General Board of Health. Further serious cholera epidemics occurred in 1848–9, 1853–4 and again in 1866 when the haphazard operation of the previous legislation was recognised. There followed the Sanitary Act of 1866, a further parliamentary commission in 1869, the new Health Act of 1872 and then the consolidating Public Health Act of 1875. The necessary legislation was now in place, but uniform enforcement was still required. Nevertheless, the overall effects of health precautions and public works, some on an unprecedented scale, in the provision of hospitals, water supplies, refuse and sewage disposal systems, were to reduce the death rate by more than a half over our period as a whole; the death rate due to cholera alone was reduced tenfold over the second half of the nineteenth century.

Public utilities

The provision of a pure and constant water supply exercised the minds of the Victorians, but this was not a new problem. There are remains of Roman dams or aqueducts in Britain, and from Saxon to Stuart times dams and ponds were built for water supply, for water-powered corn grinding and for iron furnaces and forges as well as weirs for fisheries. Plymouth, in Devon, needed water not only for its inhabitants, but also for the fleets of ships which were based there. Drake's leat, 17 miles long, 6 feet wide and 2 feet deep, brought water from the River Meavy on Dartmoor and served Plymouth for 300 years from 1591. The Devonport leat, also bringing water from Dartmoor, augmented the supply in 1793 and included several aqueducts and a 648-yard tunnel in its length. Both of these supplies were diverted into the Burrator Reservoir in 1898, but substantial sections of the leats remain. In the Georgian period, new supplies were needed to power textile mills and in addition ornamental lakes and cascades became features of the rural landscape. At Studley Royal, John Aislabie and his son produced such features on the River Skell, while the banker Henry Hoare created the landscape of Stourhead. Lancelot 'Capability' Brown built at least forty earth dams for ornamental lakes, including those at Petworth, Burghley, Harewood and Blenheim. Towards the end of the eighteenth century a new requirement for water arose: the provision of dams and storage reservoirs for canals. The first of these was built for the Arbury Canal in 1764 which served the Newdigate family's coalmines, followed by Brindley's reservoirs at Smethwick in 1771 and Smeaton's Townhead Reservoir for the Forth and Clyde Canal in 1773. The Grantham Canal, opened in 1793 and engineered by William Jessop, was the first to be entirely dependent upon stored water for its operation.[19]

Towns extracted drinking-water from local rivers which were usually polluted although the water was sometimes filtered through sand beds. Rainwater was also collected and stored and water was drawn from wells in towns and rural areas. Supplies from both were inconsistent and sometimes dangerous due to pollution; many wells were located close to cesspits into which raw sewage had been channelled. In London, supplies had first been taken from the polluted Fleet River and then spring water was brought 27 miles by conduit from Hertfordshire to Islington by the New River Company established by Hugh Myddleton in 1613. This supply eventually became inadequate and the London Bridge Waterworks utilised the River Thames, London's main sewer, as an additional source. Two water-wheels, driven by the river current, raised water which was then delivered by cart or piped through hollowed tree trunks to public hydrants. George Sorocold added a third wheel and overhauled the others in 1701. However, the Thames was often polluted, the rate of supply varied with the tide and ceased altogether in drought and freezing conditions. In 1752, a new pumping station powered by a Newcomen steam engine was built upstream at Chelsea, but this proved inadequate and cholera prevailed until additional pumping stations drew water from above the tidal limit with sand-bed filtration. The Metropolitan Water Supply Act of 1852 stipulated that all water intakes on the Thames should be above the Teddington Weir, but it took a severe outbreak of cholera the following year in areas still supplied with unfiltered water from the Thames to convince the authorities that cholera was water-borne. The requirement that a piped supply was to be available to every consumer within five years was not achieved until 1899, three years before the Metropolitan Water Board incorporated the private companies.[20] The value of aeration and sunlight in destroying bacteria naturally was established and authorities were required to provide storage reservoirs for ninety days' supply. Bleaching powder, chloride of lime, was added to water in small quantities to assist the process, but unfortunately this affected the hardness, causing a problem for steam boilers, laundries and dyeworks. The use of chlorine gas was introduced as a replacement in 1905, first in Lincoln but other water companies soon followed.

The problems of water supply were solved by Victorian engineers applying their skills to both civil engineering construction in dams and reservoirs and to mechanical engineering in pumping stations to ensure continuous supply at constant pressure. The scale of their works varied considerably from the small water-powered beam pumps such as that at Coultershaw in West Sussex, constructed in 1782, which raised water to nearby Petworth House, the home of the Earls of Egremont, to the vast schemes for the cities forced to bring water long distances from remote mountainous areas. Not all their schemes were successful. George Leather constructed the Bilberry Dam in 1839 for the Holme Reservoir Commissioners to supply Holmfirth. The earth dam, with a puddled clay core, leaked from the start and finally failed in 1852 with the loss of eighty-one lives and extensive damage which caused widespread unemployment. A similar dam at Dale Dyke near Sheffield failed twelve years later with an even worse death toll. This traditional method of dam construction was then replaced by a masonry building technique devised by William Rankine, first utilised at Lake Vyrnwy in Wales to supply the city of Liverpool. This dam, begun in 1881 and completed in 1892, was one-fifth of a mile long, 144 feet high and had a storage capacity of 12,000 million gallons. Other schemes followed with masonry dams in the Elan Valley for Birmingham, opened in 1904, and in the Lake District for Manchester, opened in 1894 – the use of Thirlmere even then provoking local environmental objections.[21]

Some water-supply schemes utilised boreholes where suitable aquifers could be tapped. Water was pumped out by various means. Examples of small animal-powered pumps may still be seen, ranging from the donkey treadwheels at Carisbrooke Castle in the Isle of Wight and Greys Court in Oxfordshire to the ubiquitous horse engines which raised water at Chilham Castle in Kent and Earlham Hall in Norfolk. Water-power was more widespread, using not only water-wheels but also turbines as, for example, at Arundel Castle where the installation was housed in a flint-faced brick building. At Sutton Poyntz in Dorset, Thomas Hawksley, one of the great Victorian water engineers, installed two turbines in 1857 to drive ram pumps and one still remains in its small stone pumphouse. On this site several phases of the water-pumping system for the Weymouth area are displayed by Southern Water in the original buildings.

More spectacular were the engine houses for the massive steam-powered Cornish beam engines which once served many water undertakings. These raised water from wells into surface storage reservoirs and water-towers by means of rods operating plunger pumps. Many of the pumping stations have architectural pretension and are, like their town halls, monuments to Victorian municipal aspirations. This applied not only to the exteriors of the engine and boiler houses, but also to their interiors and to the castings of the engines themselves. The areas surrounding were often landscaped with open cooling ponds and filter beds arranged in geometrical patterns, emulating the stately homes near which they were often located. Particularly fine tall beam engine houses, still with engines *in situ*, may be seen, that at Papplewick in Nottinghamshire being one of the best examples. Not only is the exterior ornate but the windows contain stained glass and the engine entablature columns are decorated with iron filigree work, both featuring water motifs. The two engines were built by James Watt and Co. in 1884 and they are still regularly steamed by a preservation trust. At Ryhope, near Sunderland, two engines of 1868 vintage are still steamed by a trust in the ornately gabled engine house. Two engines of similar age in the Goldstone Pumping Station at Hove are also operated, the remainder of the buildings housing the Engineerium Museum of steam power. The beam engine was replaced in the 1890s by the more efficient vertical and horizontal cylinder rotative engines. One of the earliest vertical triple expansion engines, dating from 1895, is preserved at Broomy Hill, Hereford. A very ornate engine house survives at Bratch in Staffordshire where in 1895 Bilston UDC constructed a gothic extravaganza in red brick with coloured tiles, corner minarets and crenellated parapets to contain two James Watt vertical triple expansion engines. Both of these installations required much lower buildings than beam engines.

The raw water had to be treated in open filters, following which it was stored in covered reservoirs or water-towers to prevent further pollution. In hilly areas it was possible to use gravity flow to distribute water, but in others further pumping was required to deliver treated water to all parts of an undertaking's area. The necessary pressure head for a gravity flow system could be obtained by the use of standpipes, like the 200-feet high pipe at Kew Bridge pumping station in London, which is disguised as an Italian campanile. At Broomy Hill, Hereford, there is a water-tower adjacent to the pumping station, but generally they are features of many flat rural landscapes and also of some densely-populated urban areas. Some towers echoed the style of the pumping stations with their elaborate brickwork, such as that at Finedon in Northamptonshire. Others were strictly utilitarian with iron tanks on iron, brick or stone supports, while some were completely enclosed and disguised, like the Keptie Hill water-tower at Arbroath in Scotland, built in 1885 of coarse rubble stone to resemble a battlemented castle with no hint of its actual function. Many water-towers have been replaced by modern reinforced concrete structures and an excellent comparison may be made at Rockwell Green in Somerset, where an older circular iron tank, covered by a conical lid with weather vane, stands beside the more familiar concrete type. Perhaps the most unusual of the water-towers is that at Thorpeness in Suffolk, where the tank was disguised as a house on top of a brick plinth and received its water by means of pumps driven by an adjacent windmill.

Other pumping schemes were necessary not to provide water for drinking but to drain agricultural land for cultivation and grazing. In Somerset,

East Anglia and the carrs of Nottinghamshire and Lincolnshire, there was little high ground and much of the area was so low that it could not drain naturally. The Broadlands area of Norfolk was a man-made landscape due to medieval peat digging just as the silt and peat fens are today. The land is only available for cultivation by dint of man's ingenuity in drainage schemes. These began in Roman times but major schemes to drain the fens followed the General Drainage Act of 1600 which enabled 'adventurers' to risk capital in drainage schemes in return for awards of reclaimed land. One of the first such schemes was that of the fourth Earl of Bedford who, along with thirteen co-adventurers, sought to drain the Great Level with the potential reward of 300,000 acres of peatlands. Cornelius Vermuyden, the Dutch engineer knighted by Charles I for his work on draining Hatfield Chase in Yorkshire, prepared a scheme for a system of gravity drainage by means of new river cuts, straightened channels, sluices to prevent tidal surges upstream and the creation of floodlands which could be inundated if necessary. The cuts served the dual purpose of both drainage and navigation. Vermuyden's scheme was completed by 1653 and some of the engineering works for both flood prevention and maintenance of navigation are still impressive. Denver sluice, near Kings Lynn, was originally built by Vermuyden in 1652 to prevent tidal water going up the Great Ouse. Rebuilt in 1682, 1751 and again in 1834 by Rennie, the original sluice now consists of one vertical rising floodgate, three sets of sluices and a lock with two sets of gates enabling passage of boats in either direction. A similar stone-built lock with two sets of gates was constructed at Salters Lode on the Bedford Level in 1827 at the junction of the Well Creek and the tidal Great Ouse, while a fine toll-house nearby, of local carstone and brick, is a reminder of the water-borne carriage, once the only transport in the area.

By 1700 the Vermuyden gravity drainage scheme was becoming inadequate; estuaries were silting up and as the peat dried out the field drainage channels fell below the river levels. Thus pumping became necessary and Drainage Commissions were set up, which were empowered to raise levies from landowners both to provide pumps and maintain the flood banks and drains.[22] In the eastern part of the country, the obvious source of power was the wind and eventually around 800 windmills were erected, driving large diameter scoop wheels which lifted the water about 5 feet. Subject to the vagaries of the wind, these pumps were generally successful, but were still plagued by peat shrinkage which reduced their effective lift, a problem which could be solved by either increasing the diameter of the scoop wheel or putting two or more mills in series. The flat landscapes of the fens are still punctuated by windmill towers, only a fraction of the hundreds once there, and of those extant a few have been preserved in working order. At Wicken Fen in Cambridgeshire, the National Trust protect a piece of the untouched Great Fen, now standing like an island several feet above the surrounding countryside, together with a small smock mill windpump of the type which was once commonplace in the area. A rather larger scoop wheel wind pump, erected in 1830, remains in working order at Herringfleet in Suffolk; the octagonal tarred smock mill has four plain sails and is turned into the wind by tailpole and winch. In Norfolk, along the Rivers Bure, Thurne and Yare, are several ruined brick tower windmills with scoop wheels. The tallest, at Berney Arms beside the Yare, is now maintained by English Heritage and is a landmark in the flat area around Breydon Water. This mill was built in 1840, originally to grind clinker for cement, but was converted to drainage around 1880 with the installation of a 24-feet diameter scoop wheel in a remote house – usually the wheels are immediately adjacent to the windmill tower or body. These are now but fragments of the fenland drainage schemes and the real impact of many windmills in the landscape can only be appreciated by a visit to Kinderdijk near Dordrecht in Holland.

The real answer to fen drainage lay in the steam engine which could work continuously, fuelled by coal brought along the navigable river channels. The first steam installation was completed in 1818 and by 1850 over sixty were operational. On the 25,000-acre Deeping Fen, two steam engines had by 1825 effectively drained the area formerly worked by forty-four wind pumps. One of these engines still remains at Pinchbeck Marsh, near Spalding in Lincolnshire, where the squat red-brick engine house contains an 1833 A-frame beam engine driving a 24-feet diameter scoop wheel which ceased work in 1952. At Stretham in Cambridgeshire, the 6,000-acre Waterbeach Level had been drained by three wind pumps which were replaced by a pumping engine installed in 1831 by the Butterley Company. Due to shrinkage of the peat soil, the original 29-feet diameter scoop wheel had to be replaced twice by bigger wheels in its working life. The engine continued on stand-by duty until 1957 and still remains as a monument to the steam era, but it was succeeded by Mirrlees Diesels driving centrifugal pumps and then by electric pumps. In north Nottinghamshire, the heavier silt carrs were drained by two steam engines at Misterton Soss, where only the two red-brick engine houses remain on either side of the Mother Drain beside the River Idle, once navigable up to Bawtry. On the Somerset Levels at Westonzoyland on the north bank of the River Parrett, an 1861 beam engine survives in the red-brick buildings with a tall square chimney stack which is a prominent landmark. Pumping still continues today in the fens and levels, but the more dramatic wind pumps and steam engines have been replaced by diesel and electrically-driven pumps which are available for work at the touch of a button. They do their work efficiently but their buildings contribute far less to the landscape than those of their predecessors.

The drainage of water in this rural landscape is a more attractive topic than the removal of sewage from towns, but the public health problem could not be solved without tackling the latter. As with water supply, the answer lay in civil and mechanical engineering works involving huge capital expenditure. Such was the case in London, where by 1850 there were already 1,000 miles of sewers. Even so, 200,000 cesspits remained, but in spite of this London was reputedly the best drained city in the world. In the following decade, the rise of London's population led to a vast new scheme designed by Joseph Bazalgette, with 1,300 miles of sewers and the construction of two new main outfall sewers parallel to the Thames on either bank. The existing sewers were connected to these and the effluent transported 12 miles below London Bridge to Barking Reach and Crossness on the Erith marshes. The northern outfall sewer was some 20 miles in length with three levels, ranging from aqueducts on the surface to sewers 50 feet below. The lowest level was under the new Thames Embankment, which carried below its surface a pipe subway, the sewer and the Metropolitan District Railway all running in parallel. The final lift was by pumps at Abbey Mills, West Ham, where the surviving Gothic-style engine house, with its superb interior ironwork, once contained eight beam engines arranged in pairs cruciform style. At Barking, holding reservoirs were built for the northern outfall sewer and the effluent released into the river on the ebb tide. On the south bank of the Thames, the engineering works were not as extensive, involving an intermediate pumping station at Deptford and reservoirs at Crossness, from which a pumping station raised the effluent 22 feet to be discharged into the river. This pumping station, opened in 1866, still contains the original four Watt beam engines which, although extensively modified in 1891, continued in service until about 1950. The elaborate cathedral of cast iron is the subject of a preservation scheme.[23]

Not all cities and towns were able to discharge untreated sewage directly into the sea like London. Many cities had found a ready market for the products of their night-soil collections as manure on farms, and the inhabitants frequently resented reforms which reduced this source of income. Treatment of sewage was an alternative, and sewer networks were constructed which either drained naturally to treatment plants or required additional intermediate pumps. After treatment the effluent was pumped to 'sewage farms', large areas of land purchased specifically for the purpose where underground networks of pipes with manholes enabled the effluent to be distributed on the fields. The circular artificial filters or bacterial beds with rotating distributors, a familiar landscape feature today, were introduced in the 1870s. Elaborate pumping installations still surviving include the Abbey station in Leicester, opened in 1891 with four Gimson compound rotative beam engines, and the station at Clay Mills, near Burton-on-Trent, erected six years earlier and also with Gimson engines, which pumped the massive effluent from the brewing industry there. At New Mills in Norwich, the River Wensum powered turbine-driven air compressors to operate sewage ejector pumps which pumped raw sewage to the treatment plant.

Following the improvements to the sewage system, ways still had to be found for the disposal of rubbish from towns. This could be achieved by dumping or by burning in refuse destructors or incinerators, of which there were over eighty by 1900. Installations were operating at Birmingham, Manchester and Leeds as early as 1878. Subsequent installations were improved to make use of the waste heat to provide steam for heating and also for electricity generation, the earliest plants to do this being at Ealing, Cheltenham and Oldham in the period 1895–6.[24] Electricity generation had been introduced in the middle of the nineteenth century, initially for private supply and later for public consumption. Early installations were for lighting purposes only, using carbon arc lamps and then the incandescent filament lamps devised by Edison and Swan. The first generating systems used water-power, gas and steam engines coupled to direct current generators with stand-by storage battery systems. A pioneering example of a domestic lighting supply system has been restored at the Cragside house of William Armstrong in Northumberland. In 1878 he installed a turbine to drive a Siemens dynamo, but the existing Burnfoot powerhouse dates from 1886.[25] The first public electricity supply was at Godalming in Surrey, where a mill water-wheel was connected to drive a Siemens generator in 1881. Hydroelectricity has been more generally used in Scotland, but the city of Worcester made use of the River Teme. The existing polychrome brick station at Powick was opened in 1894 and continued operation until the 1920s; three steam engines supplemented the water turbines in times of low river levels.[26]

The steam engine became the main power source to drive generators, at least until the introduction of the steam turbine by C. A. Parsons in 1884. The earliest turbine-powered public generating station was opened at Forth Banks at Newcastle-upon-Tyne in 1889, followed by another at Cambridge two years later. Power in towns was distributed by ducted underground cable systems; the grid system, using overhead cables on the pylons which are now a feature of so many landscapes, was a 1930s phenomenon. Electricity also revolutionised the provision of motive power for industry. Electric motors were introduced to drive machinery, some textile mills adapting their steam plant to generate electricity for motors which used belt drives coupled to the existing shafting. The use of electricity to power locomotives was pioneered in 1883 with the construction of railways at the Giant's Causeway in Northern Ireland and at Brighton. The latter, constructed by Magnus Volk, still operates along the sea front. Many towns already had horse tramway systems which had enabled suburbs to develop and these were converted to electric propulsion, picking up power from overhead cables. The first towns to make use of electric tramways were Blackpool in 1885 and Leeds in 1891. In London, the development of the underground railway system had highlighted the problems of air pollution from steam haulage and the City and South London Railway opened the first electric-hauled tube line in 1890. By the end of the century the electricity supply industry was only in its infancy but already the tremendous potential of the new power source was obvious. It was the most efficient way of deriving power from coal, while eventually the almost universal availability of electricity removed any restrictions on the location of factories and heralded the disappearance of many of the chimneys of the steam-power era.

The improvement in living conditions over the two centuries under consideration was by no means continuous. New industries in new locations had been obliged to provide accommodation of a reasonable standard to attract workers, many of whom had previously lived in inferior conditions. However, once the use of steam power enabled factories to be built in existing towns, the provision of housing was normally left to speculative developers and its quality deteriorated. By 1900, as a result of public health reforms, some, but by no means all, of this substandard housing had been improved or demolished. Most towns had supplies of clean drinking-water and adequate drainage, while both streets and many houses were lit by gas or in a few cases by electricity. Coal was still the ubiquitous fuel and a pall of smoke hung above the terraced houses and factories of many industrial towns. The mass production of domestic items such as pottery, glass and hardware as well as textiles had created a higher standard of material culture. The availability of urban transport, in the form of bicycles and tramways, had not only increased the mobility of the population but enabled them to live in new suburbs away from the town centres. In the country as a whole, improvements in roads and canals had transformed the carriage of bulk goods, while the national railway network was almost complete by 1900. The means by which people and goods moved around the country will be explored in the next chapter.

7

Moving around: roads, rivers, canals and railways

The period we are considering saw the most dramatic changes in methods of transport ever seen in Britain, but even in 1700 people and goods were moving around over quite long distances. People either walked, rode horses or travelled by coach while goods were carried by packhorses, carts and waggons for short distances for local distribution or by river boat to reach a wider market. London, the seat of government and the commercial heart of the nation, was then, as now, the focus of major routes. Coastal as well as international shipping brought supplies into the port of London, while John Ogilby's road atlas, the *Britannia* of 1675, portrays a web of eleven major routes centring on the capital. Some of these roads had been improved, but others were ill-maintained and often impassable in winter. Rivers, too, had been improved in the course of the seventeenth century, but navigation was often hindered by weirs for conserving fish stocks or supplying water for mills. Nevertheless, by 1700 in England and Wales only Dartmoor, Salisbury Plain and parts of the Weald, the Northamptonshire uplands, the Pennine chain and the central Welsh mountains were more than 15 miles from the sea or navigable rivers. Ports developed at the seaward ends of these rivers so that goods could proceed onward by coastal shipping.

Roads and tracks

The country was crossed by a network of tracks and roads, many of ancient origin, which were utilised for horse traffic and driving cattle. Many of these tracks centred on villages and farmsteads, giving access to their fields and upland grazing. Others covered long distances and a well-established carrier network existed by 1700. In the Pennines a system of packhorse routes had been in use for 400 years and was essential to the economic life of the north of England. Trains of small but sturdy horses, twenty to forty at a time, each carried up to 2 hundredweights of goods in panniers. Essentials such as salt, corn and fuel as well as metals, wool and cloth were carried in this way.[1] The constant wear of traffic produced sunken routes or hollow-ways, which could degenerate into natural watercourses. Where suitable stone was available, many of these were built up and paved with a single or double row of flagstones 2 to 3 feet wide and often known as causeys. If streams could not be forded, single-arch bridges were built. These were usually steeply humped to allow adequate passage for flood water beneath and had only shallow parapets which did not foul the side-slung panniers on the horses. Waymarkers were few in number until an Act of 1697 authorised local Justices of the Peace to order the erection of guideposts or stoops at remote crossroads, which pointed travellers in the right direction. One of the best-known Derbyshire causeys crosses Stanage Edge from near Hathersage and then forms the Long Causeway towards Sheffield; part of this is a paved double track to carry carts. In West Yorkshire, there is an impressive section of causey across Ilkley Moor, the town of Ilkley being a centre for packhorse routes. To the north there is a classic packhorse bridge near Thornthwaite in Nidderdale on an old track from Ilkley to Ripon. Drove roads for cattle and sheep were wider and unpaved, usually running between walls or hedges to contain the stock. The vestiges of these old road systems are still recognisable in the present upland landscape, but many lowland routes have disappeared beneath modern road construction, although Welsh Way is still named on the Ordnance Survey map just north of Buckingham.

The responsibility for maintaining both packhorse tracks and roads still lay with the parishes. This system was perhaps adequate for local traffic but not when a major through route was involved or when there were insufficient supplies of suitable road-making material. These were the problems on a section of Ermine Street in Huntingdonshire, Cambridgeshire and Hertfordshire which prompted the creation of the first turnpike trust in 1663. Three toll-gates were erected and the receipts were used by county surveyors to repair and maintain the roads and, if necessary, still to employ statute labour from the parishes. Supervision of the trust was the responsibility of the local Justices who also supervised the repair of bridges by the county, except for those in corporate towns. Seven more trusts had been authorised by 1700. Roads were improved and many of the routes which were previously two or three alternative tracks were reduced to one, the abandoned tracks becoming part of a relict landscape. Attempts to protect the road surfaces had been made by statutes limiting the number of horses per waggon and the minimum width of wheels so that the surface was rolled rather than gouged by them.

The number of turnpike trusts increased rapidly during the early years of the eighteenth century, a change in the legislation allowing the appointments of boards of trustees to manage them independently of the local Justices. The trusts were normally created for a period of twenty-one years, but sometimes less. The 1706 Act for a road from Fonthill in Bedfordshire to Stoney Stratford in Buckinghamshire, with thirty-two trustees and powers to borrow money secured against tolls, became a model for future turnpike trusts. Some of the road proposals were opposed by river navigation companies who feared loss of revenue, but turnpiking proceeded apace. In the twenty years beginning 1751, some 870 Acts were passed and by 1830 there were 1,100 trusts controlling 22,000 miles of road. Nevertheless, a further 105,000 miles of un-turnpiked road still remained in the care of the parishes. Most of the early trusts controlled only about 20 to 30 miles of road, but some, centred on towns, were responsible for a group of roads which could amount to a considerable mileage. The largest was the Hereford Trust, which in 1830 controlled 118 miles of road. Major through routes were usually controlled by several trusts, not all of which were formed at the same time. It took nearly fifty years to turnpike completely the London to Bath road westwards from Newbury and it was managed by six separate trusts.

In Scotland and Wales the road problems were worse. A programme of road-building was begun in the Highlands under General Wade in 1726, some 242 miles being built in the next ten years. Wade built the road along the Great Glen linking Fort William to Fort Augustus and Inverness, and was also responsible for the fine bridge at Aberfeldy across the River Tay. Many of these roads were suitable only for military traffic, the gradients being too great for commercial waggons and coaches. In the Lowlands, the first Turnpike Act was passed in 1714 for Midlothian, but other counties did not follow suit until the second half of the century. In Wales, too, turnpike trusts were sought by counties, first for Monmouthshire in 1755–8 and then for Glamorganshire in 1764. Abuse of the system by many of the trustees in South Wales, especially the multiplication of toll-gates, led to the Rebecca riots of 1842–3, during which many gates and toll-houses were destroyed. As a result of these riots, South Wales became the first region where roads were the responsibility of county boards. However, the vested interests of turnpike trusts prevented this happening in other places until the 1870s.[2]

Despite these improvements in the roads, many travellers still complained about their condition, none more vociferously than Arthur Young in his travels during the last three decades of the eighteenth century. In the north of England, he found the Yorkshire roads 'detestable, full of ruts whose gaping jaws threaten to swallow up any carriage less than a waggon', while the Great North Road near Darlington was 'execrably broke into holes, like an old pavement; sufficient to dislocate ones bones'.[3] Many turnpike roads were successively improved in the course of the eighteenth and nineteenth centuries so that, although some stretches are incorporated in our modern road system, others survive as a series of abandoned tracks. A good example of the latter was the Wakefield to Austerlands Turnpike over Standedge summit towards Oldham, which was first built in 1759 by Blind Jack Metcalf of Knaresborough. It included some very steep sections which were later bypassed, leaving a series of parallel tracks and abandoned loops which are still features in the barren moorland. Another eminent road engineer was Thomas Telford, who began his career building a new road system in the Highlands of Scotland. He constructed nearly 1,000 miles of new road and a similar number of bridges, ranging from small culverts to crossings of major rivers like the Dee and the Tay. He was an early exponent of the use of cast iron, for example, in the bridge at Craigellachie over the River Spey. This spanned a greater distance than the original Iron Bridge over the River Severn with half the weight of castings, which were brought by canal and sea from the Plas Kynaston ironworks near Llangollen in North Wales. His major contribution to the development of the landscape was the Holyhead Road (A5), built between 1815 and 1830 across Snowdonia. This included the magnificent suspension bridge across the Menai Straits to Anglesey and the ornate Waterloo Bridge across the Conway near Bettws-y-Coed. Telford also introduced a new method of making roads, utilising a solid foundation topped with smaller stones and a drainage culvert each side.[4] His contemporary, John McAdam, favoured a less solid foundation but a smoother road surface of rolled broken stones which bonded together under the pressure of waggon wheels. He and his sons became surveyors for 144 different turnpike trusts and 'McAdamised' a considerable proportion of nineteenth-century roads. Road improvements meant the faster transmission of mail and newspapers as well as travellers, all of which greatly speeded up the circulation of ideas and information.

Although milestones and guideposts had existed before the turnpike era, legislation in the mid-eighteenth century resulted in the trusts erecting them along their routes. Each trust had its own characteristic roadside markers, ranging from stone blocks to cast iron pillars: a good selection of the latter, painted white with black lettering and carrying the foundry name, survive on the Derbyshire turnpikes. An interesting collection of mileposts are those measured from Bow Bells in London, which are decorated with a graduated series of bells: there are several of these along the present A22 in Surrey and East Sussex. Another series of road-markers in the counties surrounding London are the Coal Tax marker-posts which were erected by the side of both roads and later railways marking the boundary, last defined in 1861, where duty was levied on coal entering London. The duty was first introduced to raise revenue to pay for the rebuilding of London after the Great Fire of 1666 and was not rescinded until 1889. The markers were in a variety of different forms, generally short or tall iron or stone obelisks and cast iron plates, many of which survive. Some restored cast iron posts may be seen in the Banstead Heath area of Surrey.[5]

A more imposing feature in the landscape were the toll-houses, built to provide living accommodation for the collectors. Many were built at road junctions, often hexagonal or round in shape with windows placed so that the toll-collector could see both ways along the road. The scale of tolls was displayed on boards, such as that by the Iron Bridge in Shropshire, which shows that not even the Royal Family were exempt from payment. Many toll-houses have been lost in subsequent road improvements, but good examples survive, such as the stone-built Steanor Bottom toll-house on the Todmorden–Rochdale Turnpike and the octagonal Butterow toll-house at Rodborough near Stroud in Gloucestershire (Plate 25). A rare surviving feature of the turnpike era is the waggon-weighing machine at Woodbridge in Suffolk: this steelyard-type machine was erected following an Act of 1741 which allowed the levy of excess tolls for overweight waggons.

Navigable waterways

The carriage of goods over long distances by turnpike could prove expensive in tolls and not all the trusts fulfilled their obligations to surface and maintain the roads. Navigable rivers still provided the best means for moving heavy goods, although some were passable only to very small boats. Improvements included dredging, bypassing shoals with cuts and the use of flash locks or staunches to allow passage through mill weirs and shallows: techniques learned from Dutch fen drainage engineers. The most sophisticated improvement was that on the River Exe in the 1560s. This bypassed the tortuous reaches of the Exe by a new 3-mile cut, involving the construction of three pound locks with mitre gates for the first time in England. Small boats carried cargoes from Exeter to Topsham, where they were transferred to sea-going vessels. The Customs House and warehouses still remain on the waterside in Exeter. A pound lock was built on the River Lea at Waltham Abbey in the 1570s with two sets of mitre gates, and more followed on the River Thames in the 1630s. The Wey Navigation in Surrey was planned by Sir Richard Weston to link the River Thames at Weybridge to Guildford. Opened in 1653, its construction involved pound locks, weirs, bridges and a 7-mile artificial channel along its 15-mile length.

Northern rivers, such as the Aire, Calder and Trent, had also been made navigable by the end of the seventeenth century in order to carry increasing amounts of coal down to the Humber where it was transshipped to coastal vessels for East Anglian ports and London. The Aire and Calder Navigation was completed in 1703, linking Leeds and Wakefield to the Humber. Other schemes in Yorkshire, such as the River Don to Sheffield

Plate 25 The attractive Butterow toll-collector's house at Rodborough, near Stroud. The octagonal house, built of local ashlar stone, is complete with its toll board showing the authorised rates. It was built *c.* 1825 for the new turnpike road from Bowbridge, on the Thames and Severn Canal, up to Rodborough.

and the Ouse to York, were completed by the 1730s, while the Calder and Hebble Navigation from Wakefield to near Halifax, engineered by John Smeaton, involved twenty-six locks and a new cut. In the south a link from the Thames at Reading to Newbury, involving eighteen locks, was established with the Kennet Navigation. This was built in the face of opposition not only from mill-owners but also from the town of Reading which feared the loss of water-borne trade. River improvements in the first half of the eighteenth century also took place in the north-west on the Mersey and Irwell from Warrington to Manchester *c.* 1725, and the Douglas Navigation from the Ribble estuary to Wigan, which was opened in 1742. The River Weaver was made navigable along the 20 miles up to Winsford in 1732, an improvement involving eleven locks but providing a valuable asset for the shipment of salt from Cheshire and for the subsequent growth of the chemical industry there.[6]

Coastwise shipping played a vital role in transport around Britain, the goods reaching the coast either by water or road. Some of the cargoes were landed upon beaches in calm weather conditions or in sheltered bays sometimes protected by breakwaters. On the north-east coast between Saltburn and Ravenscar, low tides expose a vast foreshore of relatively level rock with few stretches of sand and natural inlets. For seven months of the year these could be used for loading and off-loading cargoes and ruts were cut to guide the carts which carried ironstone or alum for shipment. However, the main proportion of shipments was handled in the river estuaries where cargoes were transferred by manhandling them across gangplanks from ship to shore. This damaged the river banks and so wooden jetties and quays were often erected. Overseas traffic was mainly to Europe and Asia around 1700 and so was concentrated at ports on the east and south coasts from Hull round to Gloucester. However, London was predominant, with nearly ten times as much ship tonnage based there as at Bristol, followed by Ipswich, Newcastle and Yarmouth. On the west coast, the smaller ports of Liverpool, Whitehaven, Glasgow and Greenock catered for the trade to Ireland and some transatlantic shipping. The safety of the coastal shipping lanes was gradually improved: lighthouses continued to be built both privately and by Trinity House, while buoys marked the shoals and sandbanks. Such was the volume of coal traffic between the north-east and London that harbours of refuge were built between the Tees and the Wash at Bridlington, Whitby and Scarborough, their construction being financed by levies on coal shipments at the port of origin.

Horse-drawn waggonways

Packhorses were employed to move cargoes of coal and other minerals down to markets or to rivers for onward shipment. The narrow tracks were unsuitable for carts and waggons, and so for carrying heavier loads, waggonways or tramways were introduced early in the seventeenth century following mining practice in Germany. These consisted of parallel wooden rails along which four-wheeled waggons were hauled by horses. The loads carried were greatly increased as one horse could pull several waggons. One of the first of these waggonways was built around 1600 by Huntingdon Beaumont for Lord Middleton of Wollaton Hall to bring coal from Strelley for sale in Nottingham. These early waggonways have left few traces in the landscape, but the courses of later ones can be identified although some have been incorporated into locomotive-hauled railways or disappeared under roads. Efforts to minimise gradients have left cut-away sections on hillsides, sometimes resembling leats for conveying water, together with embankments over valleys, tunnels and cuttings through hills. There are bridges over roads and rivers, inclined planes for scaling hillsides as well as many miles of track, sometimes with sleeper blocks and, more rarely, with rails *in situ*.

The rapid expansion of the coalfields of north-eastern England in the eighteenth century was only possible because of the use of this kind of railed transport. An extensive system of waggonways developed from the mines down to the rivers: loaded chaldron waggons, each carrying around 53 hundredweights, mostly went downhill to the rivers under gravity, the empty waggons being drawn back up hill by horses. At the riverside the coal was transferred from quays or staithes either directly into sea-going ships or into flat bottomed keels, each of which carried around 20 tons. The latter were taken down river by keelmen to deeper water where the cargo was transferred once again into the ships. The traffic grew rapidly and by 1804, 2.12 million tons, over half the total production of the area (1.5 million waggon loads), was shipped out from the Blyth, Tyne and Wear Rivers (Figure 8).

The first waggonway in the north-east was probably from Ravensworth Colliery in County Durham to the Tyne at Dunston, laid down in 1669. It has been estimated that by 1700 there were 37 miles of wooden railway on Tyneside, and 146 miles by 1800.[7] John Gibson's map, presented to Hugh, Earl and Duke of Northumberland in 1787, shows fifteen waggonways down to the River Wear on a 2-mile stretch of river. These 'Newcastle roads' or waggonways were constructed to a wide range of gauges, ranging from 3.5 feet on the Brunton to Shields line to 5 feet on the Wylam to Lemington line. The wooden rails were fixed to wooden transverse sleepers. If heavily used a second rail was attached to the first which could easily be renewed, but for even better wearing capability a wrought iron strip could be fixed on top of the wooden rail. It was, of course, often necessary for the waggonway to cross land in private ownership and the coal-owners paid an annual sum for the privilege, known as a 'wayleave'. One waggonway of which much survives is the Tanfield line, built from Dunston staithes on the Tyne to collieries at Tanfield and Causey in 1724–5. This involved bridging the Causey Burn with the famous Causey Arch, a

Figure 8 Coal drops at Sunderland from T.H. Hair, *A Series of Views of the Collieries in the Counties of Northumberland and Durham*, first published Newcastle-on-Tyne, 1844. These were originally water-colours, from which the etchings were made. Sunderland Bridge in the background was constructed high above the River Wear to allow the passage of tall-masted sailing ships. The first coal berths were built in 1812 and others followed. Coal came by waggonway from collieries inland and was lowered into the holds of ships by the drops.

stone-built bridge spanning 105 feet with a single arch which was then the longest single span in Britain. The waggonway was heavily used from the beginning, some 930 chaldron waggon-loads per day being moved along the system in 1727. Although the wooden rails, and their later replacements, have been lifted, a replica chaldron waggon and wooden railway are displayed beside the Causey Arch.

Wooden railways developed elsewhere, the north-east pattern of large waggons running on comparatively broad gauge rails being followed in Cumberland, Northern Ireland and Yorkshire. A Shropshire pattern also evolved using smaller waggons and narrower gauge rails. This was primarily for servicing the coal-mines and ironworks in the Coalbrookdale and Broseley areas, and was emulated in the West Midlands and Wales. A wooden railway was exposed during restoration work at the Bedlam furnaces beside the River Severn at Ironbridge and another at Bersham, near Wrexham. The latter probably dates from 1731 and was used to bring coal, ironstone and limestone to Wilkinson's important ironworks. The railway consisted of rectangular section timbers, ranging in width from 3.5 to 6 inches and varying in length up to about 9 feet. The gauge was about 4 feet 1 inch and the rails were pegged to sleepers set directly on the ground.[8]

The poor wearing characteristics of wooden rails have already been referred to, while the sleepers rotted or were worn away by horses. Cast iron rails, albeit in short lengths, were first introduced in Coalbrookdale in 1767. These were either pegged directly, or held by chairs, to heavy stone blocks set in a firm foundation of broken stone which better withstood the pounding from horses' hooves, but the blocks were able to move and thus the rails could go out of gauge. Some systems, such as the Brynoer waggonway in South Wales, used iron rails and transverse iron tie bars to eliminate this problem. The iron rails or plateway were originally 'L' shaped so that waggons which had plain wheels without flanges could run on them. An alternative form of track was the edge rail where a flanged wheel ran on the flat top surface of a cast or wrought iron rail supported in chairs fastened to stone sleeper blocks. This form was introduced in the north-east coalfield in the late 1790s and was the forerunner of the modern rail section. The superiority of iron rails, both in load carrying capacity and wearing capability, was soon realised: wooden waggonways were converted to use cast iron rails and additional lines built using the new material.

In Gloucestershire, an extensive waggonway system was developed during the early years of the nineteenth century to enable the exploitation of the coal and ironstone reserves of the Forest of Dean, and substantial remains of three systems may be seen. The Forest of Dean Tramroad was opened in 1809, connecting a tidal creek, Bullo Pill, on the River Severn to Soudley and Cinderford. Two stone docks were constructed at Bullo Pill and around 18 miles of plateway were laid to a 3 feet 6 inch gauge, relaid as broad gauge railway in 1854 and converted to standard gauge in 1872. The Severn and Wye Railway was opened from Lydney to Redbrook in 1810, around 25 miles in all including numerous branches. A short canal was constructed from the Severn to Lydney in 1813 and a tidal basin in 1821, making Lydney the chief port of the Forest. The main line was converted to railway, but one of the branches, the horse-drawn waggonway up the Bixslade Valley serving coal-mines and drifts as well as stone quarries, continued in use until 1947 and much of its track bed may still be seen. The third system was opened in 1812 and connected the Forest to Monmouth; the branch serving the tinplate works at Redbrook included an inclined plane which is still a feature across the Coleford to Monmouth road.[9] The main purpose of these waggonways was to transport minerals to navigable water, and the construction of canals was to lead to the laying of many more miles of waggonways.

The early canal era

The already extensive system of navigable rivers was extended from the 1740s by the addition of wholly artificial waterways or canals, reliant on local streams and rivers for water supply. Several of the early canals were promoted by individual entrepreneurs for the carriage of coal, often in a restricted area, but most major canals covered longer distances and their promotion involved a broad spectrum of investors. Companies of proprietors were created, who sought an Act of Parliament and appointed surveyors and engineers who laid out the route and prepared estimates for construction. Money was raised by the sale of shares, some of which was paid out to landowners in compensation for wayleaves across their estates. The companies hoped to recoup their capital and the running costs of the canal by levying tolls in a similar manner to the turnpike trusts. However, the absence of any form of national control meant that a disparate network of canals was created by private initiative: two widths of about 7 and 14 feet existed on different stretches of the canal network. The profit motive led to some duplication of canals, for example, those across the Pennines, and in some cases the construction of unnecessary canals which were never financially viable or, indeed, were never completed.

Canals, with their constant depth, were ideally suited for the carriage of heavy goods: 50 tons could be hauled in a canal barge by one horse compared with 30 tons in a river boat or 2 tons by waggon on a surfaced road. The earliest canals in Britain were built for the carriage of coal, the first being the Newry Canal in Northern Ireland opened in 1745 to transport coal from the Tyrone Collieries to Dublin. The improvement of the Sankey Brook in 1755 to form the St Helens Canal inspired the better-known nearby Bridgewater Canal, begun in 1761 and opened from Worsley to Manchester in 1765. Engineered by James Brindley, it incorporated underground stretches right to the coalfaces in Worsley Delph and a spectacular aqueduct across the River Irwell, now replaced by the Barton swing aqueduct. Canals solved the transport problem of land-locked coalfields, enabling them to compete for the first time with coals from the north-east brought by river and coastal shipping to London. The coalfield bridging the Nottinghamshire and Derbyshire border was able to expand because of the construction of a canal down the Erewash Valley to the River Trent: so busy did this become that a second canal followed on the opposite side of the river valley. In South Wales, the steep river valleys down to the Bristol Channel were difficult terrain for canal construction but the heavy coal traffic warranted the expense. The first of several narrow gauge canals in the area was the Glamorganshire Canal, promoted by the ironmasters of Merthyr Tydfil in 1794 to give them access to the port of Cardiff. The canal had fifty locks in its 24.5 miles' length, falling 550 feet from the Cyfarthfa ironworks to the port. Traffic soon became very heavy with 200 boats, each carrying 20 to 25 tons, taking 20 hours for the journey. Other canals soon followed in the principality, the Monmouthshire being opened in 1796 from Pontypool to Newport and the Swansea Canal being opened two years later down the Tawe Valley. The Neath Canal reached Briton Ferry in 1799 and a link was made between Brecon and the Monmouthshire Canal in 1812.[10] Such was the congestion on the heavily-locked sections of these Welsh canals down to the ports that in some cases waggonways were built alongside them. One of the first was from Penydarren to Abercynon, opened in 1802 beside the Glamorganshire Canal. Shortly afterwards it was the scene of the early trials of Richard Trevithick's first steam locomotive.

Numerous waggonways connected mines, ironworks and limestone quarries to these South Wales canals, all of which remained in isolation and were never connected to the national waterway system. The Brecknock and Abergavenny Company obtained its Act in 1793 for a 33-mile long canal from the Monmouthshire Canal at Pontymoile, near Pontypool, to Brecon. Their Act included powers to build waggonways up to 8 miles long, either by the company itself or at private expense, and several were built to coalmines, ironworks and limestone quarries. One of the most spectacular of them is Hill's Tramroad from Blaenavon Ironworks to Llanfoist Wharf, completed in 1825. From Garnddyrys forge, the ledge cut out for the original line to Blaenavon may be seen and the route to Llanfoist can be followed around the Blorenge mountain, including several cuttings and a tunnel. The waggonway descends to Llanfoist Wharf by means of three double-tracked inclined planes, where loaded waggons pulled up empties on a balanced rope system controlled by a brake. A small boat basin and warehouse remain by the canal, while the waggonway crossed the canal and descended a further incline to reach a bank of limekilns and to cross the River Usk at Abergavenny Bridge to join the railway to Llanvihangel and Hereford. The diversity of transport systems can be nowhere better appreciated than in the Clydach gorge, also connected to the canal: waggonways can be followed along both sides of the valley serving the ironworks and limestone quarries. These were plateways of 3 feet 6 inch and 4 feet 4 inch gauge, while an edge rail waggonway crossed the valley. Gravity-inclined planes which once served the limestone quarries may also be seen. The Merthyr Tydfil to Govilon road running along the north side of the river was turnpiked in 1813, but is now dwarfed by the Heads of the Valley road cut through the valley in the early 1960s. On the south side, the track of the Merthyr, Tredegar and Abergavenny Railway, a standard gauge steam-hauled line opened in 1862, may be seen threading its way through tunnels and over viaducts: this served the several banks of limekilns until its closure in 1958.[11]

In the East Midlands, an extensive system of waggonways was constructed to service the limestone quarries in north-east Leicestershire and South Derbyshire. The Ashby-de-la-Zouch Canal company was forced by lack of funds to abandon its plans to build canal links to the quarries and even to the River Trent and substituted waggonways. Extensive remains may be seen of the plateway to the Ticknall lime yards, owned by the Harpur-Crewe family of Calke Abbey, with stone sleeper blocks, some complete with oak peg and wrought iron spike which fastened the plates directly to the blocks. There are embankments, cuttings and tunnels, including one beneath the main drive to Calke Abbey, while the brick and stone bridge which carried the line over the Ticknall main street is now both a scheduled Ancient Monument and a traffic bottleneck. The main line of the waggonway to Ashby was kept open until 1913.[12]

A new era of pre-locomotive railway transport began in 1801 when an Act was passed for the construction of the Surrey Iron Railway from Frying Pan Creek on the River Thames along the Wandle Valley to Croydon, the first public line not owned by a canal company. It was later extended to serve stone quarries and chalk pits at Merstham, but traces are now largely hidden by later railway developments. This was the precursor of several other public railways, of which the first to carry passengers on a regular basis was the Oystermouth Railway, opened in 1804 originally to carry coal from the Clyne Valley to Swansea. It was later extended to Mumbles Pier and ended its life as an electric passenger railway skirting Swansea Bay, finally closing in 1960. In Cornwall, the Redruth and Chacewater Railway was built as a mineral line from Point Quay on the River Fal to serve copper- and tin-mines as far as Redruth. Originally horse-drawn, locomotives were introduced from 1854 and the line was closed in 1915. The deserted track now has a new lease of life as a long-distance footpath, linking various mining sites restored as part of the Mineral Tramways Project.

The 1820s, however, marked the watershed of the horse-drawn waggonway as experiments with steam-powered locomotives proceeded. The Act for the Stockton and Darlington Railway of 1823 authorised the company to use both horses and locomotives for haulage. The latter were used for mineral traffic from 1825, the first locomotive being Stephenson's *Locomotion No.1*, but passengers continued to be drawn by horses until 1833. Their North Road Station in Darlington is now a museum, housing this pioneering locomotive, but more typical of the line is the tiny Heighington Station, still in its original form without a platform. The first railway to convey fare-paying passengers in a steam-hauled train was the Canterbury and Whitstable line in Kent, again hauled by a Stephenson locomotive, *Invicta*, in 1830. Where steep gradients were involved, however, locomotives were not yet sufficiently powerful to cope and so hybrid railways were constructed, using stationary steam engines to convey waggons or carriages up the inclines, together with locomotives or horses on level sections. The best-known of these is probably the Cromford and High Peak Railway in Derbyshire, where one of the inclines remained in use until 1967. This railway created the first link across the south Pennines between two canals, the Cromford to the east and the Peak Forest to the west, the latter providing a necessary water-borne link to the cotton manufacturing area around Manchester. There are many surviving features on this railway, now also a long-distance footpath. These include a canal interchange wharf in Cromford where sections of the fish-bellied rails are still *in situ*, engine houses at the top of the Sheep Pasture and Middleton inclines, the latter complete with the Butterley-made steam winding engine, and a descending incline at the northern end which leads to another interchange wharf and warehouse on the Peak Forest Canal.[13] Parts of the waggonway system from the Durham coalfield to the River Tyne were incorporated in the Bowes Railway, which utilised stationary steam engines and later electric motors to wind inclines in a similar manner to the Cromford and High Peak Railway. This system remained in use by the National Coal Board until 1974, after which the section between Springwell winder and Black Fell engine passed into the care of Tyne and Wear Industrial Monuments Trust who still demonstrate incline working.

Locomotive railways

The modern railway era really began in 1830 with the opening of the Liverpool and Manchester Railway, the first to rely entirely upon the steam locomotive except for a cable-hauled incline down to Lime Street Station in Liverpool. Its construction involved extensive engineering works at the western end, including tunnels out of Liverpool into the deep Olive Mount cutting at Edge Hill where the original station building survives. Between Liverpool and Manchester, the line crossed the marshy area known as Chat Moss, which George Stephenson surmounted by cutting drains, sinking rafts of wattle and heather and then building the track on the flexible base thus created. At the eastern terminus, the original Liverpool Road Station and goods warehouse have been restored and incorporated into the Museum of Science and Industry in Manchester. The building of this railway attracted considerable public notice and a fine series of aquatint plates was published by Rudolf Ackermann in 1831 from drawings made on the spot by Thomas Talbot Bury (Figure 9). An equally fine series of drawings was made by J.C. Bourne of the next major railway construction, that from London to Birmingham in 1837–8. The magnitude of this task was redolent of the great days of Roman engineering in Britain, an endeavour reflected in the choice of classical designs for the major stations, of which Philip Hardwick's Curzon Street Station in Birmingham still serves as a reminder (Figure 10). The destruction of the great Doric arch at Euston Station, the railway gateway to the north, was a major act of vandalism which engendered new attitudes towards the conservation of the Victorian built environment.

While these railways, under the Stephensons' influence, had been built to a gauge of 4 feet 8.5 inches, a different engineer, I.K. Brunel, used the much wider gauge of 7 feet for his new Great Western Railway from London to Bristol, again pictorially recorded by J.C. Bourne. Brunel claimed that the broad gauge track gave greater stability in running, but it also greatly increased the cost of engineering works on the line. These included the bridge over the Thames at Maidenhead, which had to be sufficiently high to clear the river navigation yet not introduce unnecessary gradients on the railway. Brunel solved this problem in a daring fashion with his incredibly shallow brick arches which, although confidently expected to collapse at the time, remain in use today. Box Tunnel, near Bath, almost 2 miles long, was the greatest railway tunnel yet built when it was opened in 1841. So vast were its dimensions that wary passengers

Figure 9 Entrance of the Railway at Edge-Hill, Liverpool, from the drawings 'made on the spot' by T.T. Bury, published by Ackermann in 1831. The tunnel on the right led up to the Crown Street passenger station and that in the centre down to the docks at Wapping. The chimney stacks served the boilers for the incline winding engines which were concealed within a Moorish arch behind the viewpoint. In this sandstone cutting the trains were switched from cable haulage to locomotive for the journey onwards to Manchester. Chatsworth Street cutting, as it became known, is now disused, but was excavated for the sesquicentenary of the railway in 1980.

preferred to travel over Box Hill by post-chaise and catch the next train. Brunel's confidence in his design is indicated by the magnificent classical western portal visible from the modern A4 road. The original terminus station at Bristol at Temple Meads was equally spectacular, its late medieval style frontage intended to harmonise with the city's historic buildings. Brunel's vast trainshed, covering five broad gauge tracks with a 72-feet span, has a hammer-beam roof similar in style to that of Westminster Hall in London. The extension of the Great Western into the West Country presented further engineering problems with deep valleys and estuaries to be crossed, for which Brunel chose the cheaper solution of timber viaducts. Some of these survived into the twentieth century and their stone pillars can be seen alongside those of more modern viaducts constructed when the line was doubled, as at Moorswater. Crossing the Tamar above Plymouth required a bolder solution which was not achieved until the construction of

Figure 10 Birmingham Station from J.C. Bourne's *Drawings of the London and Birmingham Railway*, London, 1839. This was the original terminus in Curzon Street and was designed by Philip Hardwick in a classical style with a portico of four Ionic columns in front. The building contained a refreshment room with directors' apartment, secretaries' and engineers' offices. It still remains, having later been used for goods traffic.

the Royal Albert Bridge in 1859. The river is spanned by two enormous bowed oval tubes from which the track is suspended, and the railway approaches the bridge over curved viaducts at either end. This bridge was Brunel's memorial, since he died in the year of its opening and his name was emblazoned upon it. Brunel's broad gauge, however, has not survived, and was finally replaced by the cheaper standard gauge in 1892.[14]

Railway-building in the 1840s proceeded apace, so that by the time of the Great Exhibition in 1851 there were over 6,000 miles of railway in England and Wales. Its construction involved both greater expenditure of capital than canals had done and engineering works on a scale hitherto unknown because of the need to minimise gradients. This resulted in the construction of tunnels or cuttings through hills and lengthy viaducts across valleys. Robert Stephenson summed up their effects on the landscape thus:

> It seems to me but yesterday that I was engaged as an assistant in laying out the Stockton and Darlington Railway. Since then, the Liverpool and Manchester and a hundred other great works have sprung into existence. As I look back upon these stupendous undertakings accomplished in so short a time, it seems as if we have realised in our generation the fabled powers of the magician's wand. Hills have been cut down and valleys

> filled up; and when these simple expedients have not sufficed, high and magnificent viaducts have been raised and if mountains stood in the way, tunnels of unexampled magnitude have pierced them through bearing their triumphant attestation to the indomitable energy of the nation and the unrivalled skills of our artisans.[15]

Stephenson's assessment of the quality of railway engineering is attested by the survival of many of these structures in regular use today.[16] The transformation of the landscape was achieved entirely by manual labour, as is shown clearly in J.C. Bourne's lithographs of labourers at work during the construction of the London and Birmingham Railway in the late 1830s. On the same railway, Kilsby Tunnel, nearly 1.5 miles long, had to be pumped dry continuously during its construction and the spoil lifted out through vertical shafts, as in shown in dramatic fashion in several of Bourne's lithographs (Figure 11). The line of the tunnel is still clearly marked on the surface by tree-covered spoil heaps, while the shafts serve for ventilation, the two largest topped by crenellated red-brick towers. Other visible features of tunnels in the landscape are approach cuttings and portals. The Clayton Tunnel on the London to Brighton line, engineered by J.U. Rastrick in 1841, has a castellated north portal with two towers and the tunnel-keeper's cottage built on top of the entrance. The north portal of the Bramhope Tunnel, through the ridge between Leeds and Wharfedale, was built in similar style in 1849. Here enormous problems were encountered because of flooding in the 2.1-mile tunnel and twenty-three navvies died accidentally: a memorial to them resembling the portal was erected in Otley churchyard.

The Pennines presented the greatest challenge to cross-country railways as they did to canals, but the hills were eventually pierced by four lengthy tunnels. Three date from the 1840s, Woodhead, Dinting Vale and Standedge, while Totley, the longest of them, was opened in 1894 to give the Midland Railway a route from Sheffield to Manchester. Standedge had already been scaled by the turnpikes and then tunnelled through by the Huddersfield Canal before the first railway tunnel was constructed, making use of the canal tunnel for carrying materials. A second railway tunnel was added in 1871 and a third double track tunnel added in 1894, now the only one in use. At the eastern end, near Marsden, all the tunnels can be viewed from the original canal-tunnel keeper's cottage, now an interpretation centre. The wide Severn Estuary near Bristol presented an even greater tunnelling challenge for the direct line to South Wales. This was not overcome until 1885 when a tunnel extending over 4 miles was opened. Its construction was complicated by the presence of underground springs, to cope with which six steam pumping engines were installed on the Welsh side of the river at Sudbrook where only the engine house remains, now housing electric pumps.

Figure 11 Pumps for draining the Kilsby Tunnel on the London and Birmingham Railway, also by J.C. Bourne. The construction of this tunnel proved very difficult because of underground water problems and the two pumps shown in this drawing, together with a third worked by the same steam engine, raised 1,800 gallons per minute. The horse-gins were used for raising and lowering materials down the two ventilation shafts, both 60 feet in diameter, one 90 and the other 130 feet deep. The castellated red-brick ventilation towers capping these shafts may still be seen beside the A5 road.

Railways in the landscape

The railways' most impressive contribution to the landscape was the series of bridges and viaducts constructed of stone, brick, iron and concrete. The Royal Border Bridge across the Tweed at Berwick was described by L.T.C. Rolt as the most 'romantic and evocative railway structure in the world'. Constructed of stone-faced brick arches on stone piers, it curves gently 126 feet above high-water level. The Settle and Carlisle Railway was opened by the Midland Railway Co. in 1875 and includes a number of stone viaducts along its dramatic course, the best-known being that at Ribblehead. Of the many brick railway viaducts, that at Balcombe across the River Ouse in West Sussex is perhaps the most impressive. It was built as early as 1840 for the London and Brighton Railway by J.U. Rastrick: the thirty-seven brick arches are crowned by a stone balustrade and flanked at either end by Italianate pavilions. Most railway viaducts were far more functional in appearance, such as Thomas Bouch's Hownes Gill viaduct in County Durham, built to replace inclines on the old Stanhope and Tyne Railway (Plate 26). Constructed of yellow brick on very slim piers, its style contrasts with that of the plain red-brick eighty-two-arch viaduct across the Welland Valley in Northamptonshire, now sadly patched with blue brick, yet looking endless in the flat landscape.

The early use of cast iron for road bridge construction had been fraught with problems, but the stresses imposed on iron railway bridges were far more complex because of the concentrated weight and the vibration of the locomotive. Wrought iron was strong in tension compared with cast iron and was used by George Stephenson for the trusses supporting a wooden deck for his five span bridge over the River Gaunless for the Stockton and Darlington Railway in 1825. This was replaced in 1901 and is now preserved as an outdoor exhibit at the National Railway Museum at York. Subsequent early bridges used cast-iron arched ribs supported by wrought iron lattice girders and several of these may still be found in Scotland. Of particular interest are those built by Joseph Mitchell for the Inverness and Perth Junction Railway in the 1860s: the Tilt Viaduct has castellated arches at each end because of its proximity to Blair Castle, while the elegant Dalguise viaduct over the River Tay is supported by piers extended to form castellated towers. Far more ambitious was Robert Stephenson's High Level Bridge across the River Tyne at Newcastle, where the cast iron bowstring girders are supported by wrought iron tension rods and the six spans

Plate 26 The spectacular Hownes Gill Viaduct near Consett in County Durham which carried the single track Stanhope and Tyne Railway across a deep valley. It replaced two steam-hauled inclines built by Robert Stephenson up each side of the valley. The viaduct, of yellow brick with twelve arches, was designed by Thomas Bouch and erected in 1858. It is supported on slender piers which are unusual with their recessed sections, but the additional buttresses were added as a precaution after the failure of Bouch's Tay Bridge in 1879.

are carried on tall sandstone piers 146 feet above the river bed. The bridge, opened in 1849 for the York, Newcastle and Berwick Railway, still carries three rail tracks on the upper deck and a 20 feet wide roadway with two 6 feet pavements on the lower deck.

Wrought iron, being both lighter and stronger than stone or brick, enabled engineers to tackle previously insurmountable obstacles, such as bridging the Firth of Tay. Unfortunately, Sir Thomas Bouch failed to allow for wind speeds when designing the first Tay Bridge which was opened in 1878. This collapsed with considerable loss of life a year later and was not replaced until 1887. Wrought iron viaducts were also employed for spanning long distances, one of the few survivors being the Bennerley Viaduct across the Erewash Valley in Derbyshire which was opened in 1878. Robert Stephenson used two wrought iron box tubes in his Britannia Bridge over the Menai Straits which, although damaged by fire in 1970, still retains much of his original 1850 structure and now carries road as well as rail traffic. The Forth Bridge, however, was built of steel on the cantilever principle between 1882 and 1890, and remains as a spectacular engineering feat and landscape feature to the north of Edinburgh.

The use of concrete for railway structures was only introduced at the very end of the nineteenth century. The first poured mass concrete viaduct curving across the Glenfinnan Valley on the West Highland Railway is a spectacular although alien element in the Highland landscape. On the border between Cornwall and Devon, the Calstock Viaduct was built of concrete blocks at a height of 117 feet above the River Tamar and has now weathered to look like stone. Built in 1908, this was perhaps the last of the great railway viaducts as the railway network was nearing its completion.

The railway station was an entirely new feature in the landscape, although to begin with local inns occasionally served for waiting passengers. Some early stations, as on the 1832 Leicester and Swannington Railway, were built to resemble the familiar turnpike toll-houses. Other stations were specially designed to placate local landowners, such as the French château-style Boxhill and Westhumble Station on the London, Brighton and South Coast Railway, built in 'an ornamental character' to suit the owner of nearby Norbury Park.[17] A similar tiny French-style station and associated station-master's house may be found at Cromford in Derbyshire and was probably designed by G.H. Stokes, who was employed by the Duke of Devonshire at nearby Chatsworth (Plate 27). Another station designed to satisfy the whims of an individual is Monkwearmouth, built in 1848 to celebrate the election of George Hudson, 'the railway king', as Member of Parliament for Sunderland. It was built in the classical style with a huge Ionic portico and now houses a small railway museum.

Many railway companies developed their own distinctive station styles, such as the Tudor and Jacobean designs by J.W. Livock for the Northampton to Peterborough branch of the London and Birmingham Railway. Two examples remain at Oundle and Wansford, the latter now the headquarters of the preserved Nene Valley Railway. The Midland Railway favoured the use of local materials such as red and white brick, ironstone and local limestone, together with Norman-style windows with diamond or hexagonal paned windows, as can still be seen on several Northamptonshire stations. The same company employed the Nottingham architect T.C. Hine for their buildings on the Nottingham to Newark line in the 1840s. His Tudoresque station houses are characterised by scalloped barge boards and diamond-pattern latticed iron window frames; one of the best examples is at Thurgarton, now converted to a residence. The Great Northern Railway also employed Hine for their new Low Level terminus station on London Road in Nottingham, a magnificently eclectic brick building whose architectural styles range from Tudor to Venetian.

Whereas earlier railway stations were often built on greenfield sites on the outskirts of towns, the impact of later railways on established urban settlements was sometimes dramatic. At Belper in Derbyshire, George Stephenson put the North Midland Railway in a deep stone-lined cutting rather than a tunnel, which involved some demolition of property and ten overbridges in a mile to accommodate the existing streets. Further south, in Leicester, the new Great Central Line sliced through the built-up area

Plate 27 Cromford railway station, in Derbyshire, was built *c.* 1860 for the Midland Railway's extension from Ambergate to Rowsley and thence to Buxton and Manchester. The platform buildings are linked by a lattice footbridge and the station-master's house resembling a mini-château continues the French design influence. Now single-tracked, the line disappears into the tunnel north to Matlock.

in the 1890s on a viaduct, destroying nearly 500 houses whose inhabitants were rehoused in new suburbs such as Newfoundpool. A much higher viaduct dominates the Cheshire town of Stockport, with textile mills nestling beneath its tall arches. In historic towns such as Bath, York and Conway, considerable care was taken to harmonise the railway with its surroundings. In Bath, the problem had already been encountered by the Kennet and Avon Canal Company some forty years earlier, which had sunk its canal in a cutting through the Sydney Gardens and spanned it with elegant iron bridges. Brunel's Great Western Railway Company in 1840 took similar pains not to disfigure the city, with its elegant curved viaducts and bridge across the River Avon. The York and North Midland Railway encountered the problem of York city walls which they pierced with a pointed arch to give access to their first station, built in 1841. This became increasingly impractical as traffic to the north increased, and the present station, with its magnificent curving trainshed, was built outside the walls in 1877 for the North Eastern Railway. Robert Stephenson also met ancient walls at the small Welsh town of Conway and took the railway behind the castle, whose architecture was reflected in the crenellated piers of his box section river bridge: this and Telford's adjacent suspension road bridge make an elegant duo beside the thirteenth-century castle.

Similar care was frequently lavished on railway architecture in industrial towns, often as a matter of civic pride. Nikolaus Pevsner described Winton Square in Stoke-on-Trent as the 'finest piece of axial planning in the country'. It was the contribution of the North Staffordshire Railway to this important Potteries town, and the station is on the same scale as a country house, a remarkable Jacobean structure with a large square bay window above the main entrance. The remainder of the square is defined by two-storey railwaymen's houses and the matching North Stafford Hotel. Huddersfield, at the junction of three narrow valleys, had to be approached by a series of tunnels and fine masonry viaducts, all dating from the 1840s. The station was the finest classical structure in the country apart from Euston, and unlike the latter has been preserved. Its giant Corinthian portico dominates St George's Square which is also enhanced by the George Hotel, modelled on a Renaissance palace and built by the locally important Ramsden family.

The London railway termini form a ring around the city. The earliest of them, Robert Stephenson's Euston, has been sadly mauled, as also has London's second station at Waterloo. To the north of the Thames, Cubitt's King's Cross station and Scott's Midland Grand Hotel at St Pancras contrast sharply with each other. The former is the earliest surviving terminus station, built in 1852, but of surprisingly modern appearance: the plain stock brick screen with two semi-circular openings which follow the curve of the twin trainsheds behind is only relieved by the central Italianate tower. Cubitt also designed the adjacent Great Northern Hotel, whose curved façade displays a multiplicity of small windows flanked by two tiers of Venetian bays. It is a simple design compared with the high Victorian excesses of red-brick St Pancras whose fretted pinnacles pierce the skyline. The hotel forms the station façade, while behind it is the fine single-span trainshed, designed by William Barlow and the largest in the world when it was built in 1868 for the Midland Railway. Several of London's termini have been redeveloped in the late twentieth century, although most of the original façades have been retained. The most spectacular redevelopment is probably that of Liverpool Street, originally built 1874–5 for the Great Eastern Railway, where the concourse is now roofed over by a magnificent glass structure which harmonises sympathetically with the original.

As well as creating a new range of buildings, the vast railway network influenced the pattern of settlement to a much greater extent than canals had done. New housing developments sprang up alongside railways, particularly at important junctions as at Crewe in Cheshire. The early Grand Junction Railway moved their works here from Edge Hill and built a new town in the 1840s. Similarly, the Great Western company established its works at Swindon and another new railway village was created, much of which is now a Conservation Area. The county town of Derby came to be dominated by the Midland Counties Railway Co., which based its locomotive works there in 1840 and built neat terraces of cottages together with a railwaymen's club and lodging house. As well as these planned settlements, railway companies built rows of houses for their staff near junctions and depots, often in a similar company style to their station buildings. In West Yorkshire, the Midland Railway provided housing on a large scale, beginning at Normanton in the 1840s. The same company built Midland Terrace in Bradford in 1880, some forty houses next to a shunting yard. These are neat brick houses with characteristic Gothic windows picked out in polychromatic brickwork, a style the same company adopted for houses at Wigston and Syston in Leicestershire. Railway housing was often very different from contemporary local housing, the transportation of materials from elsewhere presenting no difficulty.[18]

Railways also came to influence leisure patterns, first with the building of their own hotels as, for example, at Fleetwood where the North Euston Hotel was designed to accommodate on-going boat passengers. The Zetland Hotel at Saltburn on the North Yorkshire coast was the nucleus of a new seaside resort, largely financed by the Saltburn Improvement Company, where the platform of the Stockton and Darlington Station of 1861 extended right up to the rear entrance of the hotel. The magnificent Preston Park Hotel was built in 1882 as a joint venture by two local railway companies. It was set in extensive leisure grounds overlooking the River Ribble, but was not successful and was later adapted for county offices.[19] At the other end of the social scale, the working classes of most industrial towns could reach the seaside in less than 3 hours, assisted by the provision of cheap tickets. The holiday trade was both regional and seasonal: factory workers from Leicester and Nottingham, for example, frequented Skegness, while those from the north-west went to Blackpool. Some established resorts such as Cromer resisted the social change which was threatened by the railway and preferred to rely on their genteel carriage trade, but few succeeded in stemming the tide of progress. The new trend for day excursions, pioneered by Thomas Cook in 1841, was rapidly established and the workforce was taken annually to London or to a seaside resort by special trains. Railways also changed the character of sport in the second half of the nineteenth century, transforming local matches and meetings into national events. The social influence of railways was therefore far greater than that of any previous form of transport.

Railways played an important part in the development of some ports, too, not just for taking passengers to shipping services but also for the handling of import and export cargoes. Some, dating from the horse-drawn waggonway era, were constructed as joint railway and port undertakings, for example, in South Wales at Llanelly in 1806, Porthcawl in 1825 and Saundersfoot in 1829, all mainly for the carriage of coal. The dock companies of London subscribed to the developing railways in the 1830s while railway companies involved themselves in port construction at Whitstable, Folkestone and Middlesborough. Parliament tried to restrict the involvement of railways in port development, but after 1864 they were authorised to invest in ports elsewhere and also allowed to operate steamships, providing services to Ireland, the Channel Islands and the Continent. Railway companies went on to promote large dock enterprises at Barry, Fishguard and Immingham until, by 1913, they controlled about fifty ports in England and Wales.

The canals fight back

The success of railway enterprise eventually led to the virtual demise of the canals, but to begin with it encouraged canal companies to modernise their systems. Britain's canal network had almost reached its maximum extent by the 1820s, although some schemes were never completed. Canals had enjoyed several decades of profitable trading until they began to experience competition from the burgeoning steam railways, although some enjoyed a brief prosperity carrying building materials for railway construction. Many canal companies managed to continue operations despite having to reduce their tolls or reach agreements with the railways to apportion traffic. Some improved their system in an effort to compete; others amalgamated or sold out to the railways. In the latter case, canal operations could continue under railway ownership or the canal was closed and its bed taken over for the railway permanent way.

Of the canals which sought to improve their system, one of the earliest, the Oxford, had been engineered by Brindley in the 1770s as a contour waterway between the River Thames at Oxford and the Coventry Canal. Its course maintained a constant height over long distances without cuttings, embankments or locks: building costs had been kept down but with the penalty of a long winding canal. As competition increased for traffic from the Midlands to London, journey times became more important and between 1829 and 1834 the northern section was straightened and the canal shortened by over 13 miles. This created weed-filled meanders on either side of the new line, while the new towpath was carried over the old line by elegant cast iron bridges made by the Horseley Company of Tipton, as at Braunston. At Newbold, near Rugby, a new tunnel was built, wide enough for two narrow boats to pass, and the old narrow tunnel remains beneath the churchyard. Further south at Hillmorton, the elegant canal depot is still used by British Waterways, but the locks were duplicated in order to speed traffic, as at Stoke Bruerne to the south (Plate 28).

Plate 28 The settlement at Stoke Bruerne in Northamptonshire straddles the Grand Junction Canal at the head of seven locks. The new waterway was opened between the Oxford Canal at Braunston in Northamptonshire and the River Thames at Brentford. Construction of the 100-mile long broad canal cut some 60 miles off the previous route to London via Oxford. It was begun in 1793 but was not finished until the Blisworth Tunnel just to the north was completed in 1805. Houses and shops and a steam cornmill (now a canal museum) were built beside the canal. In order to combat forthcoming railway competition the locks were duplicated in 1835–40, but commercial traffic gradually declined. In the now disused original lock on the left is a boat-weighing machine which was once used on the Glamorganshire Canal in Cardiff.

One of the last north–south canal links was the Old Grand Union Canal, opened in 1814, between Norton on the Grand Junction Canal and Foxton on the Leicester and Northampton Union. This important link through to the River Trent was only built to take narrow boats, with seven locks up at Watford and then ten locks down at Foxton, in two staircases of five. This became an impossible bottleneck and inclined plane boat lifts were planned at both places, but only one was ever constructed, at Foxton, which was opened in 1900. A massive sloping concrete ramp was built with rails upon which two steel caissons each capable of carrying one 70-ton barge or two narrow boats were hauled by means of wire ropes passing round a drum powered by a steam engine. The coal trade which had been lost to the railways never returned to the waterway and the lift was abandoned in 1911. Many of the features associated with it remain and a trust has reconstructed the engine house with the long-term aim of restoring the lift. Another important canal boat lift was that at Anderton in Cheshire which linked the Trent and Mersey Canal with the navigable River Weaver 50 feet below. Cargoes, particularly salt, had been transshipped at Anderton by chutes and tramways, but in 1875 the twin caisson lift was opened. Originally the caissons were raised and lowered hydraulically using steam power but corrosion of the hydraulic rams required a major modification in 1908. A new framework carried pulleys over which wire ropes attached to the caissons and balance weights passed, lifting power being provided by electric motors. Once again corrosion has taken its toll and the lift, now a scheduled Ancient Monument, again requires drastic restoration.

One of the most heavily used canal systems in nineteenth-century Britain was in the Birmingham area, where by 1906 the Birmingham Canal Navigations (BCN) operated 159 miles of canal with 216 locks. This system had continued to develop after the 1850s in parallel with railways in the area, some twenty-six railway interchange basins being built. At Smethwick, the urban canal landscape is particularly evocative, with remnants of the original canal and two subsequent improvements. In 1769, Brindley had proposed to tunnel through the hill but eventually scaled it by six narrow locks on either side to the summit. These soon became a source of congestion and twenty years later Smeaton constructed a new canal using cuttings to eliminate three locks on each side, reducing the summit level. The final improvement was carried out by Telford in 1829, and cut through the hill in a deep cutting, the new lockless summit level being built for barges and having double towpaths. Telford's Smethwick cutting is crossed by several bridges, including the single-arched cast iron Engine Aqueduct, Telford's handsome 150-feet span cast iron Galton Bridge, the blue-brick Steward Aqueduct which carries the original canal to Oldbury and Dudley and most recently the M5 motorway on stilts.

The Engine Arm Aqueduct is a reminder of the great problems the BCN system had with water supply. It sought to solve these difficulties by means of storage reservoirs and pumping systems to return water to the summit levels. This navigable feeder brought water from a pumping station at Bridge Street, Smethwick, the site of which has recently been excavated and is open to the public. The Boulton and Watt engine which worked here from 1779 to 1897 was re-erected at the Birmingham Museum of Science and Industry in 1983 where it is regularly steamed. The red-brick pumping station which replaced this used gas engines and stands beside Telford's new main line, from which it lifted water to Smeaton's 473 feet level which still remains in use.[20] At Windmill End, Netherton, on the Dudley Canal, a junction remains among the landscaped tips of coalmines and ironworks, some connected by their own canal arm with elegant cast iron bridges to carry the towpath. This is one of the most evocative relict industrial landscapes in Britain and will be discussed in more detail in Chapter 8.

In order to fight the railways on their own ground at the height of railway mania, the Ellesmere and Chester Canals amalgamated with the Birmingham and Liverpool Canal to promote the Shropshire Union Railways and Canal Company, which obtained three Acts to build railways. The London and North Western Railway countered this threat by leasing the canals in perpetuity and later incorporating the Shrewsbury, Montgomeryshire and Shropshire Canals into the Shropshire Union. This company operated a large boat fleet and provided a service integrated with the railways. The Shropshire Union canals are well endowed with landscape features, their hilly terrain being conquered by inclines and aqueducts pioneering the use of cast iron. There were steam-wound tub boat inclines on both the Shrewsbury and Shropshire Canals. The incline at Hay achieves a vertical rise of 207 feet in a horizontal distance of about 1,000 feet, the equivalent of around twenty-seven average locks, and the incline may still be seen at the Blists Hill Open Air Museum. The Shrewsbury Canal was opened in 1797 to connect that town to the coal- and ironworks at Donnington Wood. The original engineer, Josiah Clowes, built a masonry aqueduct to carry the canal over the River Tern but it collapsed during flooding in 1795. Thomas Telford replaced Clowes as engineer and built an iron aqueduct, 186 feet long on piers some 16 feet above the river at Longdon-on-Tern: the canal is now closed but the aqueduct remains in splendid isolation. This was the prototype for the much more ambitious iron aqueduct across the River Dee at Llangollen, originally planned as a three-arch stone structure with locks at either end to reduce its height above the river. The aqueduct, completed in 1805, is over 1,000 feet long and rests on eighteen tapering stone pillars some 126 feet above the Dee. It is one of the most spectacular feats of canal engineering. William Jessop, the principal engineer for the Llangollen branch, had also to cross the River Ceiriog at Chirk. Here he kept to stone construction but used a cast iron liner for the ten-arched aqueduct which is approached from the Welsh side by a 459-yard tunnel. The arrival of the parallel railway viaduct carrying the Shrewsbury to Chester line adds to the sense of grandeur.[21]

Association with railway companies was not the panacea for the continued survival of all canals. The canal from Melton Mowbray in Leicestershire to Oakham had been opened in 1802, only to be sold to the Midland Railway for their Syston to Peterborough branch in 1847. The new railway used much of the canal line but in places chose a more direct route, leaving some of the contour canal as a relict landscape. Some sections remain in water, as near Ashwell, while at Market Overton the wharf buildings remain beside the dried up canal, much of the line having disappeared under the plough. Further south, the Kennet and Avon Canal also finished up in railway ownership. It had provided a link between the Thames at Reading, via the River Kennet, to the Avon at Bath. Within 57 miles there were seventy-nine broad locks, twenty-nine of them in the magnificent Caen flight rising 237 feet near Devizes where side ponds were constructed to conserve water. Water for the canal in this chalk area was always a problem and supplies were obtained near the summit at Crofton and at the western end at Claverton near Bath. The former, a steam-powered pumping station, used two early Watt beam engines to raise water from Wilton Water. The engines have been restored by the Canal Trust and are regularly steamed. The station at Claverton, designed by John Rennie, has also been restored. This used the River Avon to power a water-wheel which drove beam-operated plunger pumps to lift water from the river into the canal 53 feet above. At the western end of the canal there are two impressive aqueducts which carry it over the River Avon, one at Avoncliff and the other at Limpley Stoke. The latter, named after Charles Dundas, the chairman of the canal company, was modified to bridge the Great Western Railway from London to Bristol. This opened in 1841 and canal revenues were soon affected both by competition and by continuing water shortage. In 1852 the canal was sold to the Great Western which continued to maintain it, but by 1909 only local traffic was using the canal and abandonment was finally proposed in 1950. The Kennet and Avon Canal Trust was then formed with the aim of restoring the canal for pleasure boating. This was achieved in 1990 with the restoration of the Caen flight, but through traffic is still plagued with the old problem of water shortage.

Canal-building on a larger scale took place during the nineteenth century with the construction of ship-canals designed either as short cuts for sailing vessels or to avoid hazardous river estuaries. In Scotland, canal-building was initiated by the government as part of their policy to open up the Highlands after the 1745 Jacobite Rebellion. The first and smaller ship-canal was the Crinan from Ardrishaig on Loch Fyne to Crinan on the west coast, designed to eliminate the lengthy passage round the Mull of Kintyre. Begun in 1793, it suffered from a shortage of labour during the Napoleonic Wars and was not really fit for use until 1809. It was designed to take vessels 88 feet long by 20 feet wide along its 9 mile course, and was mostly used by fishing vessels until a steamer service was opened from Glasgow to Fort William in 1828. Much of the trade to the Western Isles from Glasgow passed through the canal on steam 'puffers', but traffic now consists mainly of sailing yachts. The Caledonian Canal through the Great Glen was a far larger undertaking, providing a coast to coast waterway from Fort William to Inverness. Surveyed first by James Watt and then by John Rennie, it was finally completed by Telford in 1822. The 60-mile long route, of which only 21.5 miles was actually canal and the remainder large natural lochs, could accommodate vessels 160 feet long by 36 feet wide, but their height was unlimited because swing-bridges were provided throughout. There were twenty-nine locks, mostly grouped into staircases to save expense in construction. The most spectacular sight on the canal is Neptune's Staircase at Banavie, where eight locks raise the canal up from sea level at the western end. The canal is crossed here by the Road to the Isles and the West Highland Railway to Mallaig, both on swing bridges.

The Gloucester and Sharpness Canal was built to avoid the tortuous lower reaches of the River Severn. It took nearly thirty-five years to complete because of financial problems and never reached its original destination at Berkeley Pill. When completed in 1827, the 16-mile long level canal was 90 feet wide and could accommodate 750-ton ships. At the original entrance a sea lock gave access to a tidal basin from which broad and narrow locks transferred vessels on to the canal itself. This and the lonely dock-master's house are now disused, since by 1874 increasing traffic and larger ships required the construction of a new sea lock, tidal basin and dock to the south. There were numerous warehouses here, particularly for the corn trade, although all but one of these have been replaced by concrete grain silos. Another important item of trade was coal from the Forest of Dean, brought to Sharpness by a railway constructed across the Severn in 1879. The abandoned pillars and abutments of the swing-bridge section over the canal still remain, although the rest of the bridge over the river was destroyed after two vessels collided with it in 1960. The canal crossed the Stroudwater Navigation at Saul by a level crossing where the Junction House remains. At the Gloucester end of the canal is the most complete set of warehouses remaining in Britain, picturesquely grouped against the backdrop of the cathedral. With their rectangular outline, small windows and vertical lines of loading doors, they typify the transit warehouse of the period, in which imported goods were stored prior to transshipment via the Severn and canal network to the Midlands. Some were later converted into flour mills following the introduction of roller milling, and the City Flour Mills still function. The other warehouses have been restored for a variety of purposes, the North Warehouse serving as the local council offices and the Llanthony Warehouse of 1873 providing a fine setting for the National Waterways Museum (Plate 29).[22]

Plate 29 The basin at Gloucester Docks, constructed in 1812 at the head of the Gloucester and Sharpness Canal, which eventually terminated at Sharpness when opened in 1827. Boat traffic could enter the River Severn for travel upstream in through a lock in the far left background. A fine range of warehouses survives, many of them built to store corn imports and most of them now being reused. These include the North Warehouse (1827), now the offices of Gloucester City Council in the background beside the domestic-style dock offices. On the right bank are several large multi-storey warehouse blocks. The largest of them, in the right foreground, is the former Llanthony Warehouse, built in 1873 for corn merchants, which now contains the National Waterways Museum. Beside it a traditional canal maintenance yard has been created.

The greatest canal undertaking in Britain was the construction of the Manchester Ship Canal which was finally opened in 1894. The city of Manchester was anxious for a waterway connection because of Liverpool's excessive port charges, but Liverpool was equally fearful that the Ship Canal would increase problems of silting in the Mersey estuary. This was the first canal to make use of steam shovels and dredgers in its construction, together with the labour of up to 17,000 men. There were three locks at Eastham into the Mersey, enabling 600-feet long vessels to enter the canal, while other locks were duplicated to allow for different width vessels as well as having intermediate sets of gates to save water when shorter vessels were passing through them. The Barton Swing Aqueduct was completed in 1893 to allow Britain's oldest major canal, the Bridgewater, to pass over this her newest one. Docks were established at Manchester and Salford, which became busy inland ports for a comparatively short period since the Ship Canal, despite its spectacular crossing by the M6 over the Thelwall Viaduct, is now disused and in danger of closure.

Docks and harbours

The purpose of many canals and railways was to transport goods shipped in and out of seaports, some of which developed rapidly as a result of increased overseas trade. Shoreline quays and jetties with moorings subject to tides and weather were replaced by closed dock systems in which water could be retained at all states of the tide. The earliest dock facilities were those provided for the Royal Navy and these Royal Dockyards, financed and operated by the government, were responsible for the construction and repair of most of Britain's warships. The first had been established by Henry VII in 1496 at Portsmouth, while the next phase of building looked to the Continent with dockyards on the Thames at Woolwich (1512) and Deptford (1513), followed by Chatham on the River Medway in 1570. The Anglo-Dutch Wars of the mid-seventeenth century saw further Royal Dockyards established at Harwich and Sheerness, followed by Plymouth in 1691. The Royal Dockyards were further expanded during the Napoleonic Wars and continued to serve the Royal Navy through two world wars until, with changing methods of warfare, they have been gradually run down. Fine examples of covered shipbuilding slips remain at Sheerness and Chatham. At the latter, many Georgian buildings are open to the public including the 1,140-feet long red-brick rope walk where traditional rope-laying is still carried on. At Portsmouth, in addition to historic ships, many of the Georgian and Victorian buildings of the naval dockyard are still in use, albeit not for their original purpose. Also in Hampshire, at Gosport, is the Royal Clarence Victualling Yard, formerly Weevil Yard, with its remaining collection of brick and sandstone buildings designed to produce food for the navy; these included a bakery making ships biscuits, a brewery and cooperage. The later Royal William Victualling Yard at Plymouth was completed in 1832: designed by John Rennie, it was built of Devonian sandstone with granite detailing. This fine group of industrial buildings was built on a 14-acre levelled site set around a square ship basin on the rocky Cremill peninsula.[23]

Bristol, second only to London in terms of ship ownership in 1702, stood at the junction of the Frome and Avon Rivers, and moorings on the riverside quays were subject to tidal fluctuations. Trade in the port, based both on re-exports and the slave trade with Africa and America, increased considerably during the eighteenth century and the port became very congested. Several schemes of limited success were replaced by William Jessop's Floating Harbour. This entailed damming the River Avon at Rowsham and diverting it through the New Cut, thus allowing the old river courses to be maintained at permanent high water with entrance locks at the western end and a further entrance for small vessels via the Bathurst Basin from the New Cut. This ambitious scheme was completed between 1804 and 1809, but was subject to silting, a problem later tackled by I.K.

Brunel who raised the Netham weir to enable the basin to be scoured regularly. The sizes of the entrance locks were also increased, but trade in the port gradually declined as larger steamships found the passage up the Avon Gorge difficult. New docks were built at the mouth of the Avon and trade into Bristol itself largely ceased. However, the basic form of the Floating Harbour may still be seen.

The most enduring symbol of the Port of Bristol is the suspension bridge over the Avon Gorge at Clifton, designed by I.K. Brunel and completed in 1864 after his death. Vessels passed beneath it and through the Cumberland Basin into the Floating Harbour. The engine house which provided hydraulic power for operating lock gates and capstans remains with its accumulator tower, and is now used as a public house. Around the harbour itself, many of the original buildings and installations have been cleared and replaced by new housing and industrial developments, although the 1880s engineering workshops built by the port authority, together with a second hydraulic power station, still survive. The original Great Western Dock now contains the *Great Britain*, the steamship launched in 1843 by Brunel and now undergoing restoration. On the north bank, by Princes Street Bridge, the fine Bush tea warehouse built in the 1830s remains a distinctive landmark. Between Princes and Bristol Bridges, more warehouses survive, rising straight up from the water's edge. On Welsh Back, set away from the quayside, is a former granary built in 1869 in what is known as 'Bristol Byzantine' style; the same Moorish features are echoed by new housing beside the Bathurst Basin and the remains of an 1875 oil-seed warehouse. Dominating the western end of the harbour are three massive red-brick bonded tobacco warehouses which were built in the first decade of the twentieth century.[24]

Liverpool was a comparatively new port compared to Bristol yet participated in the same lucrative African and American trade, the two ports together threatening London's dominant position. Ships had to be beached upon the shallow Mersey foreshore until the first 4-acre dock was opened in 1715, Britain's first dock combining legal and commercial quays. It was Jesse Hartley, appointed engineer to the port authority in 1824, who has left his imprint upon the dock landscape of Liverpool. In order to maximise shoreline space Hartley built enclosing walls well out into the estuary. Clarence Dock was built especially for steamships, to be followed by Brunswick Dock in 1832 and then by three more totalling 29 acres between 1834 and 1836. Probably the finest dockyard complex in Britain is Hartley's Albert Dock opened in 1845, the first in Liverpool with bonded warehouses. The 8-acre basin is almost completely enclosed by fortress-like five-storey red-brick warehouses. The front walls are built flush with the quay edge supported by massive cast iron pillars to leave open space beneath, while the upper storeys have loading doors and cranes which permit direct working between the warehouse and ship. Most of the hydraulically-powered hoists and cranes have been preserved, but the dockyard was really too small when it was built and finally closed in 1972. At Stanley Dock further north, Hartley provided a link to the Leeds and Liverpool Canal in 1848. Two five-storey warehouses are dwarfed by the huge thirteen-storey bonded tobacco warehouse built in red and blue brick in 1900 on the partly infilled dock. Nearby the castellated hydraulic accumulator tower and imposing entrance gates provide a fine example of Hartley's use of random granite stone for building.

As the size of ships increased the square basins surrounded by warehouses became inconvenient and the branched dock layout was introduced, incorporating several long but relatively narrow quays opening up from a larger basin; ships could be worked on both sides of the quay using cranes. There were no warehouses, goods being taken elsewhere for storage either by road or railway. The first docks of this type in Liverpool were Langton (1879) and Alexandra (1880). Congestion on the dockyard railway system was alleviated by the construction of an overhead railway which opened in 1893. The 7-mile system with fourteen stations was built on an iron viaduct supported by iron pillars, in many cases duplicating existing surface railway lines. This system closed in 1956 and few traces of it remain other than pillars embedded in buildings.[25]

The River Clyde also shared in the transatlantic trade, with goods being landed at Greenock and Port Glasgow before the development of a dock system upriver in Glasgow itself. Extensive enclosed docks were constructed in Greenock at intervals throughout the nineteenth century, together with multi-storey dockside warehouses. The dredging and straightening of the Clyde enabled dock development to take place in Glasgow itself from the 1860s with the opening of the Kingston and Queen's Docks. Unusually, multi-storey warehouses were not a feature of Glasgow Docks, goods being stored temporarily in transit sheds before loading or discharge. Goods were held in long-term bonded and other warehouses in the area surrounding the docks or in the city itself, which boasts some fine examples of nineteenth-century warehouses, notably the classical buildings in James Watt Street. Clydebank was also the home of many important shipbuilding companies whose wealth is reflected in many of Glasgow's public buildings.[26]

London was the dominant port during our period, but until the construction of docks ships lay out in the Thames and goods were transferred to lighters before being unloaded at the legal quay on the north bank upstream of Tower Bridge and sufferance quays located dowstream on both sides of the river. The first proper dock on the Thames was the 10-acre Howland Dock at Rotherhithe, built between 1697 and 1700 expressly for empty ships to lay idle rather than for handling cargoes. Congestion grew in the port, but it was not until the last decade of the eighteenth century that cargo docks were considered. Two closed docks were built, the first being the West India Docks of 54 acres, opened in 1802 and entered from Blackwall Reach. The second was the single 20-acre London Dock at Wapping for North American and European trade, opened in 1805 and defended by 20-feet high walls. The City of London financed the construction of a canal across the Isle of Dogs which shortened the river passage for ships to the Pool and to the Wapping Dock. Near the Tower, St Katharine's Dock, opened in 1828, consisted of two small 4-acre docks with an entrance basin and was suitable only for sailing ships. Originally enclosed by tall warehouses, some of which still remain among replica modern buildings, this dock was dwarfed by later dock construction for London. The 100-acre Victoria Dock, when opened in 1855, was the largest in Britain with an entrance lock 80 feet wide and quays half a mile long. It was devoid of warehouses, all goods being removed by railway to the City warehouses 4 miles away. New construction and improvements of London's docks continued to accommodate larger ships, but, as at Bristol, new docks began to be built down river where there was deeper water and more space for development.[27] A 450-acre dock estate was developed at Tilbury 26 miles away, beginning in 1886, and slowly London itself ceased to be a major port. Bristol, too, had to give way to Avonmouth and Liverpool to Seaforth as ships of a size never dreamed of in the mid-nineteenth century were unable to penetrate their river estuaries and had to remain near deeper water.

The transport revolution in the two centuries we are considering was perhaps the most far-reaching of all the changes which took place. Although people and goods had been moving around in 1700, it was comparatively slowly by packhorse, horse-drawn waggon or river boat and coastal sailing ship. Larger ships bringing in goods from overseas were generally either beached for unloading or moored in mid-river and their cargo transferred into smaller boats. A century and a half later, roads had been improved to such an extent that a fast coach took only 42 hours to travel from London to Edinburgh compared with at least ten days before. The regularity of the mail coaches established a timetable which came to govern business transactions. Navigable rivers had been supplemented by canals, both major cross-country ones like the Kennet and Avon and complex networks in heavily industrialised areas such as the West Midlands. In the remaining half-century, both roads and canals, previously supplemented rather than superseded by the horse-drawn waggonway, lost out to the far faster steam railway, which revolutionised the carriage of both people and goods: the journey time between London and Edinburgh was further reduced to under 9 hours. At the ports, the transshipment of cargoes was no longer subject to tidal conditions as sea-going ships could be moored in enclosed docks where continuous working was possible. Many of these were dominated by the massive warehouses provided for the interim storage of goods.

The transport revolution had many effects on the economy, not least of which was to help redress the imbalance of both population and wealth between the north and south of the country. Much of the country's mineral wealth lay in the north and west, areas which had previously been difficult of access, but the strides made in civil engineering during this period enabled the physical barriers to be overcome and the coalfields of the north, the Midlands and South Wales developed into major industrialised areas. Changes in the pattern of overseas trade, with America and the West Indies challenging the domination of Europe, led to the decline of east coast ports such as Ipswich and to the spectacular growth of Bristol, Liverpool and Clydeside with their new enclosed docks and transit warehouses.

In the landscape, the construction of canals and even more of railways modified the topography of the country, thrusting straight across the land rather than following the natural contours. Staircases of locks carried canals across the Pennine watershed, while the railways spanned deep valleys and wide estuaries on impressive viaducts. In the towns, railways had equally drastic effects: large areas of land were sterilised to serve as marshalling yards for rolling stock, while cuttings and embankments created new barriers in built-up areas. On the other hand, the railways contributed magnificent new buildings to the townscape in the form of railway stations and hotels. They also pioneered new techniques of building for the airy trainsheds over station platforms and the large workshops necessary for locomotive and carriage construction. In their turn, many of these have become redundant and long stretches of derelict canal and railway can be traced in the landscape, now serving as peaceful havens for wildlife while modern traffic hurtles down miles of motorways. The landscape has been permanently transformed by man's need to move around and is itself a palimpsest of the ways in which he has chosen to do so.

8

The industrial landscape: past, present and future

Images of industry

This book has been concerned with the different facets of the industrial landscape in the two centuries between 1700 and 1900. Industrial activity was widespread during this period and until the second half of the nineteenth century was integrated with, rather than superimposed on, the landscape. The attitudes of contemporaries to the vast changes that were taking place are recorded in the diaries of many travellers throughout Great Britain during the eighteenth and, more especially, the nineteenth centuries. These were mainly members of the lesser gentry, clergymen and professional people who emulated their betters on the 'grand tour' by exploring the sights to be seen in their own country. In the growing romanticism of the late eighteenth century, they viewed waterfalls, mountains and ruins but could not fail to be impressed by the increasing industrial activity they found. Some came deliberately to seek it out. These were the so-called 'industrial spies' from Europe, who came to discover the secrets of Britain's early industrialisation. Their numbers included men with specialist knowledge of the metallurgical industries, like the Swedish travellers Svedenstierna, Schroderstierna and Triewald and the Frenchmen Gabriel Jars, Le Turc, Moissenet and de la Houlière.[1] Their accounts were largely factual and some, like Jars and Moissenet, made meticulous drawings of the equipment they saw. Others went so far as to smuggle out dismantled machines so that they could be copied in their home countries.

Many English travellers were equally interested in the novel processes which they witnessed, from Defoe in the early eighteenth century to semi-official reporters like Arthur Young and John Farey later on, together with romantically-minded clergymen like the Reverend Richard Warner and the Reverend James Plumptre.[2] Others, like William Cobbett and Viscount Torrington, bewailed the desecration of the landscape by industry, yet could not fail to be impressed by the scale of what they saw. Torrington purchased a pair of coarse cotton stockings in Askrigg in 1792 but decried the construction of a cotton mill by the bridge across the Ure at Aysgarth:

> Sr. Rd. Arkwright may have introduced much wealth into his family, and into the country; but, as a tourist, I execrate his schemes, which, having crept into every pastoral vale, have destroy'd the course and beauty of Nature; why, here now is a great flaring mill, whose back stream has drawn off half the water of the falls above the bridge.[3]

In the Derwent Valley of Derbyshire, he found that Arkwright's pioneering mill at Cromford had replaced the 'rural cot' and 'the stream perverted from its course by sluices, and aqueducts, will no longer ripple and cascade'. However, Torrington was impressed by the cotton mills in nearby Wirksworth and, from a naval family himself, felt they resembled 'a first rate man of war; and when they are lighted up, on a dark night, look most luminously beautiful'.[4] Seven years earlier, this same image had captivated Joseph Wright of Derby, who painted Cromford Mill at night with all its windows aglow, a sight never previously witnessed in rural England.

Artists and writers together convey images of industrial activity in such a manner as to express contemporary attitudes in a way not possible by other means. The windmill was a popular subject with many artists of the Romantic period because it dominated the rural landscape at the same time as drawing attention to the changing nature of the sky. Constable's summer holidays in Brighton enabled him to paint West Blatchington windmill, a unique structure in which the mill body is mounted upon the roof of a barn (see Plate 5, p. 24). Several artists, notably J.S. Cotman, were fascinated by the windmill constructed within the Gothic ruins of St Benet's Abbey in Norfolk, whereas Britton, the antiquarian, stated in a matter of fact manner that 'the remains of the once stately building are no more, except part of the magnificent gateway, and this is partially obscured by a draining mill erected over it'.[5] A similar contrast in attitude can be seen between the bald, factual description of the Merthyr Tydfil ironworks by the Swedish ironmaster Svedenstierna and the striking depiction of them by Thomas Hornor (see Figure 5, p. 91). Svedenstierna lists the 'three blast furnaces, three refineries, twenty-five puddling- and eight bloom-furnaces, with the necessary hammers and rolling-mills, as well as nine or ten steam engines, some of which act with the strength of 70 to 80 horses'.[6] Hornor emphasises the scale of the ironworks through the devices of exaggerated perspective and the play of light and dark in the sky above, reminiscent of a modern laser light display.

The intense fiery glow of ironworks excited the imagination of those only used to candlelight and oil lamps, as is depicted in de Loutherbourg's famous paintings of the Bedlam and Madeley furnaces in Coalbrookdale, the red and yellow light creating a spectacular backdrop to the shadowy industrial scene. Visitors to ironworks often felt they were venturing into unknown infernal regions, as did Webb in his tour of Ferreday's ironworks at Bilston in 1810–11:

> Plain narrative is inadequate to convey what I felt at this wonderful combination of the ingenious productions of man! I was not only astonished by the works of art, but Nature had also contributed to add terror to the scene, by the earth smoking at different places; in consequence of the burning coal-pits, by which these works are surrounded. . . . The roaring of the furnaces, the clanking of iron-chains and machinery reminded the traveller of the poetical descriptions given us of the infernal regions. . . . The awfulness of the scene reminded me of the description of the Cyclops forging thunder for Jupiter.[7]

Other travellers recorded their excited horror at the way in which mining ripped the landscape apart, nowhere better seen than at the Parys Mountain copper-mine on the Isle of Anglesey. J. 'Warwick' Smith's water-colour of the mine, painted about 1790, depicts the great chasm referred to by the Reverend Bingley who visited the area not long afterwards:

> I stepped on one of the stages suspended over the edge of the steep, and the prospect was dreadful. The number of caverns at different heights along the sides; the broken and irregular masses of rock which everywhere presented themselves; the multitudes of men at work in different parts, and apparently in the most perilous situations; the motions of the whimseys and the raising and lowering of the buckets to draw out the ore and the rubbish; the noise of picking the ore from the rock, and of hammering the wadding when it was about to be blasted; with at intervals the roar of the blasts in distant parts of the mine, altogether excited the most sublime ideas, intermixed however with sensations of Terror.[8]

Even today, the landscape of Parys Mountain, with its range of coloured rock and limpid turquoise pools is an awe-inspiring spectacle presided over by a derelict windmill tower and steam engine house which once pumped water from the great opencast.

Contemporaries had to become accustomed to the impressive scale of mills, factories and civil engineering structures, especially during the nineteenth century when canals and railways advanced across the landscape. Black's tourist guide of 1850 regarded the railway between Chester and Holyhead as 'rendered pre-eminently remarkable by those stupendous and wonderful triumphs of modern engineering, the Conway and Britannia tubular bridges'.[9] The same guidebook also describes J.U. Rastrick's Balcombe Viaduct as 'the most stupendous work of its kind in the kingdom', a view shared by G. Earp Junior in the aquatint he made in 1841 to commemorate the opening throughout of the London and Brighton railway. Various artistic devices were used to monumentalise these structures, showing them in disproportionate scale to the surrounding features or omitting the latter altogether. The reverse was also sometimes favoured, the engineering structure being submerged within an exaggerated landscape to emphasise the sheer scale of the undertaking. Edwin Dolby's lithograph of the South Eastern Railway depicts a tiny train emerging from the Abbotscliff Tunnel, while Spreat's lithograph of the South Devon Railway near Dawlish shows the railway threading its way through tunnels cut in highly exaggerated cliffs. For other artists, however, these man-made structures were easily absorbed into the rural landscape, with cows and sheep shown grazing unperturbed beneath viaducts and families viewing distant industrial scenes. Tait's *Views on the Manchester & Leeds Railway* of 1845 are a fine example of this attitude towards the integrated landscape, with country people up on the moors gazing at the distant mill towns of Brighouse, Halifax and Wakefield and the railway itself forming a minor feature in his pictures.

The life of the workers in those mills is perhaps better documented than that of most of the working class, because the government set up various commissions to investigate the labour problems created by this new form of organised production. It is difficult to appreciate the degree of bias in these reports, but they do at least allow the working man and woman to speak for themselves. Their working methods can also sometimes be gleaned from the day-to-day records maintained by their employers in the course of business, such as day books, letter books and ledgers, and the bare bones of these can be fleshed out by using paintings and engravings, particularly in the cotton industry where the toils of the child-piecers and machine-minders attracted public sympathy. The plight of women and children in the coalmines was highlighted in the Mines Commission's reports of 1842 with its heart-rending sketches of young children sitting in dark tunnels minding the air doors or women carrying baskets of coal up ladders to the surface. However, not all illustrations of working scenes were undertaken to highlight social problems; some were done to explain processes, like the simple engravings in Tomlinson's *Cyclopaedia*, while others were spontaneous records of mundane tasks, like J.C. Bourne's sketches of masons and other craftsmen building the Paddington Tunnel. The same artist also depicted the railway navvies overcoming immense physical obstacles by manual labour, for example, during the excavation of Tring cutting on the London and Birmingham Railway. The human dimension within the industrial landscape can be appreciated from these various contemporary sources.

Reconstructing industrial landscapes

All these observations indicate the vast scale of industrialisation in Britain before 1900. In the twentieth century, the nature of industry has altered, and this change has often led to the destruction of many earlier industrial landscapes. Only in the last decades of the century has the importance of these been realised, as the part Britain played in world industrialisation has been increasingly appreciated. On the positive side, recent legislation has tried to ensure that adequate recording and research are done before irrevocable actions are taken. Recording and survey work has been carried out in many regions, for example, on textile mills in northern England and elsewhere, but this has covered but a fraction of the total number. Many landscapes and buildings remain in urgent need of study, and this book has tried to demonstrate the interrelationship of the various elements which make up an industrial landscape. A textile mill, for example, can only function with a supply of raw materials, a power source, means of production, a workforce and a system of distribution to its market. These separate elements must each be examined in order to understand their contribution to the whole. In any industrial landscape, the elements enumerated in Chapter 1 which should be considered are as follows: sources of raw materials, processing plant, power sources, secondary industry, accommodation and transport. Many elements also change through time – a steam engine, for example, replacing a water-wheel and then itself being supplanted by a gas engine or electric motor. All the elements are not always present, and some which have left no trace in the landscape can only be determined from documentary evidence. The ideal method is to examine the surviving physical evidence, to try to ascertain any anomalies or missing elements in the present landscape, and to use map and documentary evidence to supply the deficiencies where this is possible. One is therefore using the present landscape as the framework on which to reconstruct the past.

This can be illustrated by looking at the evocative landscape of Windmill End in the Black Country, on the southern slope of the ridge crowned by Dudley Castle (Plate 30). The empty shell of an engine house dominates a tranquil scene in which a length of canal disappears into a distant tunnel and branches lead off on either side, each crossed by elegant cast iron tow-path bridges. A blue-brick bridge crosses the canal before the tunnel and the engine house is surrounded by graded waste tips. These are the only surviving remnants of a thriving industrial community founded on the local raw materials of coal, clay and ironstone. From at least 1700 the area had specialised in small iron goods, with Netherton as a centre, and the manufacture of nails and spades was the foundation upon which later ironworks were to be based. The first Dudley Canal, opened in 1779, passed to the west of Netherton. It was not until 1798, when the Dudley No.2 canal linked the first to the Worcester and Birmingham Canal, that the Windmill End area had the transport element essential to its future development. Such was the traffic congestion on the Dudley canals that in 1858 Netherton Tunnel was built, the last major canal tunnel to be built in Britain. This linked Windmill End directly to the Birmingham Canal Navigation, giving it three routes to the national network. At the same time, a straight cut bypassed the Bumble Hole loop giving direct access to the tunnel, although the loop continued to give access to various canal basins until the end of the century. The area's transport connections were further improved with the opening of the Dudley to Halesowen railway in 1878, together with a branch to the Withymoor Basin which acted as a canal–rail interchange and enabled vast quantities of coal, iron goods and pottery to be dispatched (Figure 12).

The development of the canal network in the area, together with the availability of steam power, transformed the industrial landscape. Lack of water-power had prevented earlier large-scale exploitation, but once steam pumping and winding engines and blowing engines for blast furnaces were available the early surface workings for coal and ironstone could go deeper and several new ironworks were built in the area. The processing plant was established, and the iron industry progressed from smallware manufacture to integrated ironworks producing wrought iron for structural purposes and heavy objects such as anchors. For example, the Netherton Furnaces, originally established before 1812 by the Earl of Dudley, were by 1873 working four blast furnaces, together with forty-two puddling furnaces and three rolling mills. There were several other groups of furnaces, including those at Withymoor and Windmill End which had both been demolished

Plate 30 An industrial landscape transformed. A canal crossroads in the Black Country just south of Netherton Tunnel near Dudley. The black and white bridges were cast at the Toll End Works in Tipton and were erected in 1858 when the new tunnel was opened. Cobb's engine house, belonging to the former Windmill End Colliery, dominates the scene.

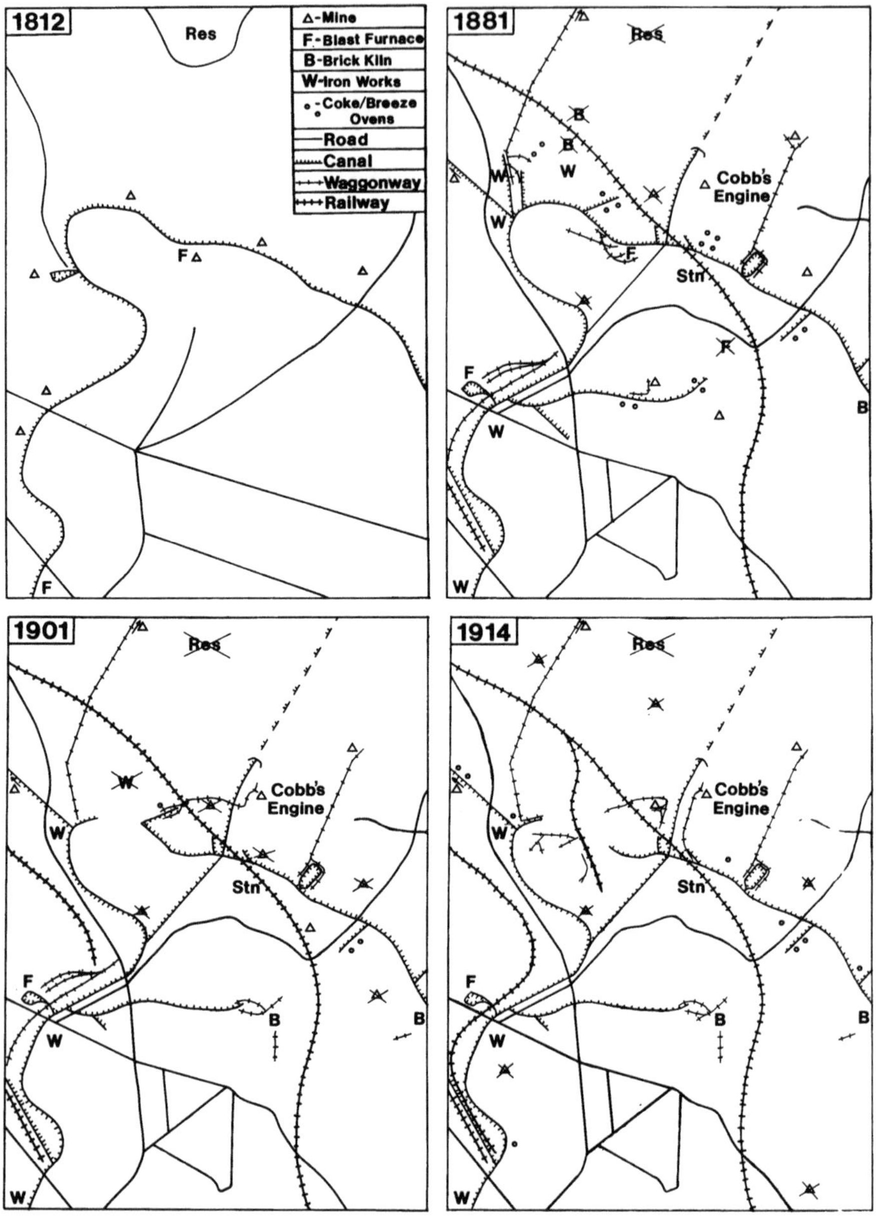

Figure 12 The transformation of an industrial landscape, Windmill End, near Dudley, from 1812 to 1914. The information is derived from the 'Plan of the mines of Lord Dudley and others', dated 1812 and the first three editions of the 25 inch to 1 mile Ordnance Survey maps. The symbol is crossed through when the particular plant is shown as 'disused' at the date of survey.

by the date of the survey for the second edition of the Ordnance Survey 25 inch map in 1901, whereas Netherton continued working throughout the First World War. Secondary iron industries were established to utilise local iron, including boiler-making, and the spade, shovel, nail and chain works continued on a larger scale.

Larger collieries replaced the scattered shafts of the previous era, most with direct waggonway or incline access to the canal. Many of them built batteries of coke ovens to provide fuel for the furnaces, and some of the waggonways were used for tipping both coal waste and red-hot furnace slag. The scene may be recreated from the writings of contemporaries, such as Elihu Burritt, reporting to the American government in 1868, who contrasted the Black Country and its 'Green Borderland':

> It matters not which you see first; whether you dip into this district of fire and smoke and artificial thunder and lightning from the greenest and quietest of rural landscapes, or into these from the black forest of forge and furnace chimneys; each produces a sensation of mind from the contrast, which it would not if seen by itself alone.

He also commented on the subsidence of land and buildings which he regarded as a characteristic feature of the Black Country:

> Right under those terrible furnaces the moles are at work night and day rooting out walks through deep coal-seams. Under the foundations of tall-steepled churches all a-light with the evening lamps and resounding with the voices of devotion, the pickmen are at work grubbing lanes under towns, hills, railways and canals. Everybody seems to feel that they live, labour, eat, and sleep on a very uncertain and unsteady footing. But the decline is very gentle. A house seldom if ever sinks so deep that its occupants have to escape through the roof.[10]

Accommodation followed the normal pattern in coalmining areas of scattered hamlets grouped around an established road pattern. By the mid-nineteenth century mining and iron-making had reached its peak and few changes can be detected in the housing pattern up to 1900 from map evidence.

The colliery owners in the area amalgamated their pumping systems in 1870, continuing to use engines built earlier in the century, such as Cobb's engine on the No. 3 shaft of the Windmill End Colliery. It is the red-brick house for this which still survives as a scheduled Ancient Monument, and it once contained a single acting pumping engine built about 1831. This remained in operation until 1928, when the engine was scrapped, but its companion winding engine, an old fashioned atmospheric engine, has ended its life in the Henry Ford Museum in Detroit. The photograph from the late nineteenth century (Plate 31) must be compared with the present day (Plate 30) to appreciate the amount of clearance that has taken place

Plate 31 Windmill End Colliery in the late nineteenth century. Cobb's pumping engine house can be seen on the left with a horse capstan in front which was used for raising pump rods from the shaft. The atmospheric winding engine is on the right, which wound from two shafts. Steam for this engine was supplied from the haystack boiler on the extreme right.

since closure. Only waste tips and ponds caused by subsidence indicate the presence of several other collieries nearby. Similarly, there are no remains of the several brickworks which were established to utilise the local clays, although the former Doulton's Birmingham Pottery, manufacturing sanitary ware, still operate to the east. All these products found an outlet along the canal, together with the ragstone from nearby Rowley Regis. The sources of raw material and processing plant were therefore interdependent with the transport system at Windmill End. This complex landscape of nineteenth-century industry can only be recreated by the use of maps, engravings, photographs and documents, yet the landscape itself provides many clues to the past.

Preserving the past

The destruction of the nineteenth-century landscape is not unique to Windmill End. The first half of the twentieth century saw rapid development, particularly in urban areas, which altered or destroyed many important industrial sites. But in the closing decade of the twentieth century, many would argue that Britain has now entered a post-industrial phase. Heavy industry based on the extraction of raw materials, with the exception of road-stone and gravel, has largely disappeared, together with the large-scale processing of natural fibres previously used in the textile industry. Foreign imports and new materials have taken their place, bought on a world market from areas with cheaper labour costs. New technology and greater efficiency of production have led to the concentration of manufacturing in fewer locations with a far-reduced labour force. The transport infrastructure which played so important a role in Britain's industrial development has been run down: canals have been allowed to decay and the railway network drastically curtailed. Consequently, the pace of change has slowed down in older industrial areas, many of which are regarded merely as derelict land, their buildings demolished and the areas landscaped, sometimes following the burial of contaminated waste materials from previous use. Regrettably, previously untouched greenfield sites are becoming new centres of industry and invaded by characterless new factories and distribution depots whose locations are governed by motorway access rather than sources of power or raw materials.

However, the international significance of many of Britain's older industrial landscapes and buildings is being increasingly recognised, and the techniques used for their conservation and interpretation adopted as a model in other countries. An analysis of the statutory protection which can be afforded to buildings and landscapes is beyond the scope of this book, but the most important are scheduled as Ancient Monuments and more may well be treated in this way during the next decade.[11] Buildings whose quality meet the criteria of both historical significance and aesthetic appeal are likely to be listed, while groups of buildings may have been designated as Conservation Areas. Statutory protection does not, of course, always ensure a building's maintenance or guarantee public access, and so many voluntary groups have formed charitable trusts to maintain and interpret local buildings, particularly those still containing machinery which can be demonstrated to visitors. This kind of conservation clearly depends on local enthusiasm and does not exist in all parts of the country. Local authorities and National Parks have now begun to recognise the tourist potential of industrial landscapes, particularly mining landscapes, in their care and to provide trails and low-key interpretation material.

Where industrial structures or buildings cannot be retained in their original setting they have sometimes been carefully dismantled and rebuilt on a new site, often forming part of a collection, such as those at St Fagans near Cardiff or at the Avoncroft Museum in Hereford and Worcestershire. Alternatively, they can be moved into a sympathetic industrial setting where some monuments already exist, such as the Black Country Museum at Dudley or the North of England Open Air Museum at Beamish in County Durham. One outstanding area which embraces all these facets of conservation is the Ironbridge Gorge, which is a museum within a landscape whose importance was recognised by its designation as a World Heritage site in 1986. Central to the Museum is the Iron Bridge itself, erected by Abraham Darby across the gorge of the River Severn and opened to traffic in 1781 (Plate 32). This was the first major bridge in the world to be built of cast iron and its survival is remarkable. However, structural problems caused by movement of the stone abutments required urgent major action in the 1960s and led to the foundation of the Ironbridge Gorge Museum Trust which supervised the work. The Trust also took under its care the original blast furnace which produced the iron for the bridge members. A charcoal furnace dating from 1658, it was adapted in 1709 for Abraham Darby's experiments with coke smelting and rebuilt again in 1777 to cope with the massive production of iron required for the bridge. It had miraculously survived under debris and, despite the fact that it is now housed in a modern protective structure, its original context can still be observed, with a storage pond for its water supply which was one of several down the dale. The ironworks were managed by the Coalbrookdale Company, a paternalistic organisation which constructed rows of housing for its workforce, several of which still survive, along with the managers' houses, in the vicinity of the furnace. The company produced iron castings and the storage warehouse for these, with its fine cast iron clock, now contains the Museum of Iron. The whole Coalbrookdale area forms one of the visitor complexes within the Ironbridge Gorge Museum.[12]

Another ornate warehouse alongside the river contained goods for shipment, and has been utilised as a Museum of the River Severn. The trows going downstream would have passed the Bedlam furnaces, the glow from which when they were working is recaptured in the red and gold of de Loutherbourg's famous painting. The present structures are the only survivors of a group of purpose-built coke blast furnaces dating from the 1750s, representing the second stage in the development of iron-making in the Severn Gorge. The ruined stone-built furnaces have been left as elements in the landscape, whereas further down river the Trust has preserved intact a pottery and an encaustic tileworks in which the techniques of manufacture are demonstrated. The final element in the Ironbridge Gorge Museum is the 42-acre Blists Hill Open Air Museum, which attempts to recreate a Shropshire coalfield village in the last years of the nineteenth century. Certain elements were already present on the site, including a tub boat canal with an inclined plane down to the river, a small mine which supplied clay to brickworks across the canal and the remains of both plateways and standard gauge railways. The main surviving feature was the remains of the Blists Hill ironworks begun in the 1830s by the Madeley Wood Company. In contrast to the other blast furnaces, these were iron-cased and blown by large vertical steam engines, one on either side of the

Plate 32 The Iron Bridge across the Severn Gorge in Shropshire, the world's first iron bridge cast in Abraham Darby's Coalbrookdale Foundry in 1777 and erected in 1779. It is now the central feature of a World Heritage site in the care of the Ironbridge Gorge Museum Trust (Photograph by courtesy of the Ironbridge Gorge Museum Trust).

three furnaces. This enterprise closed in 1912, and only the stone furnace bases and engine houses survive, a great loss since there are no preserved iron-cased furnaces in Great Britain. The adjacent ironworks was also cleared, but has been replaced by puddling furnaces, a steam hammer and rolling mills, mainly from Walmsley's Atlas Forge in Bolton, the last works in the world to make wrought iron, which closed in 1976. This plant is contained within a re-erected building, designed by John Rennie in 1815 which was brought from Woolwich Dockyard. The complex is occasionally operated to produce wrought iron bars by traditional methods and is but one of several old craft techniques which are demonstrated in authentic re-erected buildings, brought on to the site and grouped to present a Victorian streetscape complete with gas lighting. The visitor can change his pocket money into old pennies, step back in time and spend them in the sweet-shop, buy candles or enjoy a pie and pint in the village pub with sawdust on its floor. The Ironbridge Gorge Museum therefore attempts to combine archaeological integrity in the treatment of its world-famous monuments with the provision of an enjoyable visitor experience, since the funding of the former largely depends on the income from the latter. Its future evolution, like that of many similar museums, is dependent not on an individual's first visit but on the provision of sufficient new attractions to tempt the visitor back time and again.

The large numbers of museums created in the last decade, particularly those concerned with past industry or Victorian social conditions, has led to the criticism that Britain is wallowing in nostalgia for its past greatness now that its industrial lead has passed to other countries.[13] There is a limit to the number of museums which can be accommodated, as has been shown by the failure of many to sustain their vital visitor numbers, and other ways of conserving the industrial heritage have to be considered. The past development of the landscape, as was shown in the first chapter of this book, reveals that change is brought about by the interaction between topographical factors and human needs. The latter have changed considerably in the late twentieth century and the landscape once again has to accommodate them, albeit with due environmental safeguards. Industrial sites have always been adapted: many water-powered sites, for example, started life as corn mills, then went on to fulling cloth, forging or slitting iron and making paper. A striking example of this is Guns Mills in the Forest of Dean, which began life as a charcoal blast furnace at the time of the Civil War but had a paper mill built on top of it in the mid-eighteenth century (Plate 33). Some mills and factories were converted to alternative production with minimal alteration: Yorkshire woollen mills adapted well to cotton production, while in the East Midlands hosiery factories lent themselves easily to boot and shoe manufacture.

Adaptive reuse

Not every industry has created a legacy of reusable buildings. Many of the surface buildings associated with extractive industries were ephemeral, often constructed of wood and corrugated iron, while other structures such as blast furnaces, steel rolling mills, tinplate works and chemical plants do not lend themselves to adaptive reuse. It is in the food-processing, manufacturing and warehousing sectors of industry that the most durable buildings are to be found. In order to withstand the vibrations and weight of machinery, these buildings had to be substantially constructed, often of fire-resistant materials, with regular window patterns to give natural light. When originally built, many mills and factories were the largest secular structures in a community and as such have come to be regarded as local landmarks. Those dating from the water-powered phase of production are often in attractive settings, as are mills and warehouses alongside canals and docks. All of these factors emphasise the potential of reusing industrial buildings, an argument put forward by SAVE Britain's Heritage since the mid-1970s.[14]

The most easily adaptable of all industrial buildings are those in rural

Plate 33 A problem for reuse – Guns Mills in the Forest of Dean, part scheduled Ancient Monument and part listed building. The seventeenth-century charcoal iron furnace produced armaments during the Civil War. In the mid-eighteenth century, the ironworks was converted to a paper mill which used the furnace stack as a stairwell and cut doorways through it. A large timber-framed building was placed on top of the furnace, but paper-making ceased in 1879. An application for conversion to residential use has long been under consideration.

situations, such as water-mills, malt-houses and small river and canal warehouses. Many are converted to dwellings or restaurants, while modern technology has enabled many offices to operate from such rural locations. The attractive limestone-built water-mills of the former county of Rutland have made attractive houses without destroying their character, for example, at Ketton, Tickencote and Tinwell. Even larger mills, such as those at Aston Cantlow in Warwickshire or Coxe's Lock beside the Wey Navigation in Surrey (Plate 34), have been successfully converted to flats. Similarly, brick-built maltings in Oundle and Southwell are now homes, whereas the attractive flint malt-house, with its distinctive kilns, beside the navigation at Stowmarket has become a public house. The best-known of all maltings conversions is probably at Snape, near Aldeburgh in Suffolk, home of the Benjamin Britten music festival. The attractive location and style of canal warehouses also lend themselves to conversion, either into inns as at Shardlow on the Trent and Mersey or offices as at Marple on the Peak Forest. Windmills, almost always in rural settings, by their very form do not lend themselves to easy conversion which inevitably entails the removal of the internal machinery. However, many have been utilised for living accommodation while retaining their sails, but new adjoining buildings detract from their former solitary setting. So fashionable have such dwellings become that the reader may be deceived by totally new buildings in the windmill, maltings or warehouse idiom.

The conversion of large industrial structures to alternative use presents a whole new range of problems. These include high capital costs, which can often delay the completion of a project and leave parts of the structure unfinished. The conversion of both the striking Bliss Tweed Mill near Chipping Norton, with its candlestick chimney, and the long range of Dunkirk woollen mills at Nailsworth have been affected by this problem. The demand for housing within easy commuting distance of London has been the salvation of many Thames-side warehouses or large East Anglian maltings, originally built beside railways as at Long Melford and Sawbridgeworth. No such solution has been found for the fine range of listed maltings at Sleaford in Lincolnshire or the former cotton spinning mill at Cressbrook in Derbyshire, both of which are too isolated. Such is the scale of the problem of redundant textile mills in the north of England that comprehensive surveys have been made in order to determine priorities for their retention and reuse.[15]

The sheer scale of many industrial complexes has prevented their adaptation, as in the case of the railway works in Crewe and the engineering works in Newcastle-upon-Tyne. The Great Northern Railway warehouse in Derby is still standing derelict and ambitious plans for a leisure park await an upturn in the economy. Local initiative has been responsible for other successful projects, such as the conversion of the former Dean Clough carpet mills in Halifax to an industrial park containing over 200

Plate 34 Coxe's Lock Mill on the Wey Navigation in Surrey. Granaries and flour mills continued to operate here, using grain brought by barge from London Docks, until the late 1960s. Luxury conversion projects like this are in considerable demand in south-east England.

small firms. In nearby Saltaire, the vast mill complex has been turned over to multiple uses which include an art gallery, the whole benefiting from its setting within a model community. A similar solution has been found for Liverpool's Albert Dock, which now houses part of the Tate Gallery as well as a maritime museum, housing and a shopping complex. Warehouses at Gloucester Docks have also found multiple uses as council offices, museums and retail outlets. These are not always successful, as can be seen at London's Tobacco Dock, too far from the city centre to be commercially viable. Occasionally only the façades of large buildings can be preserved with new functional structures behind them, as in the cases of the Egyptian-style former Kayser–Bondor hosiery factory in Baldock and Green Park Station in Bath, which both front superstores.

Groups of industrial buildings, often with their associated housing, may have been incorporated into conservation areas. Most of the early textile communities, such as New Lanark in Scotland, Cromford and Darley Abbey in Derbyshire and Saltaire on the outskirts of Bradford, are protected in this way. So, too, are examples of model housing like the railway village at Swindon or the Voysey-designed colliery housing at Whitwood in West Yorkshire. Many Victorian towns have equally important integrated developments of red-brick terraced houses, workshops and factories, such as the estates developed in Kettering in Northamptonshire by the shoe manufacturers themselves. These reflect the various phases of development in the shoe industry over three decades of the nineteenth century and they are every bit as important as the landscapes of the early textile industry, yet are not afforded the same statutory protection. Industrial conservation areas in towns are more difficult to achieve than those in rural locations, principally because modern industry requires access for road vehicles and space for car parking rather than proximity to railway or canal. However, some enlightened city councils such as Nottingham have encouraged the reuse of Victorian factories and brought about the rebirth of some run-down industrial areas. In the suburbs of Radford and Basford, selective demolition of factories, or the conversion of ground floors, has provided car parking and access, while the renovation or replacement of adjacent substandard housing has restored living communities to these areas. Similar treatment has ensured the continued use of the warehouses of Nottingham's Lace Market and Bradford's worsted warehouses in 'Little Germany'. The Victorian townscape has therefore been retained, yet functions as a living entity.

Despite the vast scale of change during the twentieth century, today's industrial activity is not entirely unconscious of its past. Although modern transport has reduced the importance of the local availability of raw materials and electricity eliminated the need for proximity to water-power or coal supplies, the existence of a skilled labour force often dictates that a particular industrial activity remains in the same place. It is still important that lace is made in Nottingham, wool and worsted around Bradford and cutlery in Sheffield: the historical location of industry is now an important marketing factor which has undoubtedly helped prevent its wholesale destruction. Buildings in many industries have been adapted to suit modern methods and although museums of industry have proliferated to an alarming extent over the past decade, the industrial past has not been totally fossilised: the historical process of change in the landscape has continued.

In many areas, however, particularly in extractive industries where the evidence of past workings are landscape features rather than buildings, this process of change has often resulted in total clearance because of modern environmental concerns. However, even in such areas, local authorities are now interested in retaining links with their industrial past and buildings or structures may well be consolidated as monuments, like Cobb's engine house at Windmill End. These are the pointers to the past, and it is hoped that the reader of this book has been enabled to recognise their significance and to visualise their previous environment. A methodology for taking apart the layers of the past has been suggested, and the authors trust that the reader will enjoy the challenge which this presents and so come to understand the immense importance of Britain's industrial heritage.

Notes

1 The location of industry in the landscape

1 Thirsk, Joan, *The Rural Economy of England*, London, The Hambledon Press, 1984, pp. 217–57.
2 Defoe, Daniel, *A Tour Through the Whole of England*, vol. 2, London, J.M. Dent, 1927, letter VII, p. 89.
3 Lindsay, Jean, *A History of the North Wales Slate Industry*, Newton Abbot, David & Charles, 1974, p. 88.
4 Ward, J.T., 'Landowners and Mining', in J.T. Ward and R.G. Wilson (eds), *Land and Industry*, Newton Abbot, David & Charles, 1971, p. 70.
5 Palmer, Marilyn, *The Richest in All Wales*, British Mining No. 22, Sheffield, Northern Mine Research Society, 1983.
6 Keast, John, *The King of Mid-Cornwall: The Life of Joseph Thomas Treffry (1782–1859)*, Redruth, Dyllansow Truran, 1982, pp. 122–36, 145–53.
7 Letter from Lord Rawdon to Mr Coltman, 20 December 1786, Huntington Library MSS. HA5125; letter from Lord Rawdon to Earl Ferrers, 1793, Huntington Library MSS. HA5128.
8 Biddle, Gordon, *The Railway Surveyors*, London, Ian Allan, 1990, pp. 92–5.
9 PRO. Rail 803/l, Ashby Canal Minute Book, letter Lord Rawdon to Bakewell, 31 August 1792; letter Lord Rawdon to Ellis Pestell, 2 September 1792.
10 Biddle, op. cit., p. 81.
11 Biddle, Gordon and Nock, O.S., *The Railway Heritage of Britain*, London, Michael Joseph, 1983, p. 224.
12 Simmons, Jack, *The Railway in Town and Country 1830–1914*, Newton Abbot, David & Charles, 1986, p. 308.
13 Trinder, Barrie, *The Industrial Revolution in Shropshire*, Chichester, Phillimore, 1973, pp. 319–34.
14 Mee, Graham, *Aristocratic Enterprise*, Glasgow, Blackie, 1975, p. 141.
15 Tangye, Michael, *Tehidy and the Bassets*, Redruth, Dyllansow Truran, 1984, p. 48.
16 Young, Arthur, *A Six Months Tour through the North of England*, vol. III,

1771, reprinted New York, Augustus Kelley, 1967, p. 132.
17 ibid.: vol. I, p. 49.

2 *Providing the necessities of life*

1 Simond, Louis, *An American in Regency England*, ed. Christopher Hibbert, London, Pergamon Press, 1968, p. 103.
2 Gregory, Roy, 'The Use of Power in the Early Industrial Development of Hull', *Industrial Archaeology Review*, vol. XV, no. 1, Autumn 1992, pp. 7–20.
3 Conway-Jones, Hugh, *Gloucester Docks: An Illustrated History*, Gloucester, Alan Sutton, 1984, pp. 50–9.
4 Coad, Jonathan G., *The Royal Dockyards 1690–1850*, Aldershot, Scolar Press, 1989, pp. 287–8.
5 Wright, Neil R., *Lincolnshire Towns and Industry 1700–1914*, Lincoln, History of Lincolnshire Committee, 1982, pp. 180–3, 189–92.
6 Ashmore, Owen, *The Industrial Archaeology of North West England*, Manchester, Manchester University Press, 1982, pp. 10–12.
7 Buchanan, C.A. and R.A., *The Batsford Guide to the Industrial Archaeology of Central Southern England*, London, Batsford, 1981, pp. 161, 171.
8 Owen, Colin, *The Greatest Brewery in the World; A History of Bass, Ratcliff and Gretton*, Chesterfield, Derbyshire Record Society, 1992.
9 Jones, Gwen and Bell, John, *Oasthouses in Sussex and Kent: Their History and Development*, Chichester, Phillimore, 1992, pp. 24–45.
10 Prestidge, Stanley (ed.), *The Call to Duty*, London, HMSO, 1983, pp. 17–20.

3 *Fuel and power for industry*

1 Hart, Cyril, *The Industrial History of Dean*, Newton Abbot, David & Charles, 1971, pp. 61–104.
2 Harris, Helen, *The Industrial Archaeology of Dartmoor*, Newton Abbot, David & Charles, 1972, pp. 98–115.
3 Nef, J.U., *The Rise of the British Coal Industry*, London, Routledge, 1932, pp. 19–22.
4 Harris, J.R., 'The Employment of Steam Power in the 18th century', *History*, vol. LII, no. 175, 1967, pp. 133–48.
5 Farey, John, *General View of the Agriculture and Minerals of Derbyshire*, 1811–17, reprinted Matlock, Peak District Mines Historical Society, 1989, p. 338.
6 Palmer, Marilyn and Neaverson, Peter, 'The Steam Engines at Glyn Pits, Pontypool: An Archaeological Investigation', *Industrial Archaeology Review*, vol. XIII, no. 1, Autumn 1990, pp. 7–34.
7 Chapman, Nigel A., 'Ventilation of Mines', *Industrial Archaeology Review*, vol. XV, no. 1, Autumn 1992, pp. 45–58.
8 Young, Arthur, *A Six Months Tour through the North of England*, vol. II, 1771, reprinted New York, Augustus Kelley, 1967, p. 261.
9 Svedenstierna, Eric T., *Svedenstierna's Tour of Great Britain 1802–3*, ed. M.W. Flinn, Newton Abbot, David & Charles, 1973, p. 72.
10 Hey, David, *Yorkshire from AD 1000*, London, Longman, 1986, p. 277.
11 Jackson, Stephen (ed.), *Industrial Colonies and Communities*, Lancaster, CORAL, 1988, pp. 43–9.

12 Caffyn, Lucy, *Workers' Housing in West Yorkshire 1750–1920*, London, HMSO, 1986, pp. 107–26.
13 Lewis, E.D., *The Rhondda Valley*, London, Phoenix House, 1959, p. 202.
14 Atkinson, Frank, *The Great Northern Coalfield 1700–1900*, Newcastle upon Tyne, Graham, 1966, pp. 59–62.
15 Battye, Kay, Doncaster, Richard, Mitchell, Ian and Newing, Don, 'Summerley Coke Ovens', *Derbyshire Archaeological Journal*, vol. 109, 1989, pp. 180–93.
16 Atkinson, Frank, *Industrial Archaeology of North-East England*, Newton Abbot, David & Charles, 1974, p. 47.

4 Metals in the service of man

1 Lewis, P.R. and Jones, G.B.D., 'Dolaucothi', *Antiquaries Journal*, vol. XLIX, 1969, pp. 242–72.
2 Rieuwerts, J.H., 'A List of Soughs in the Derbyshire Leadmines', *Bulletin of Peak District Mines Historical Society*, vol. 3, no. 1, 1966, pp. 1–42.
3 Turnbull, Les, *The History of Lead Mining in the North East of England*, Alnwick, Sandhill Press, 1987, p. 53.
4 Palmer, Marilyn, *The Richest in All Wales*, British Mining No. 22, Sheffield, Northern Mine Research Society, 1983, p. 30.
5 Rogers, K.H., *The Newcomen Engine in the West of England*, Bradford on Avon, Moonraker Press, 1976, pp. 16–27.
6 Barton, D.B., *The Cornish Beam Engine*, Truro, D. Bradford Barton, 1966, pp. 22–7.
7 Willies, L.M., Roche, V.S., Worley, N.E. and Ford, T.D., *The History of Magpie Mine*, Matlock, Peak District Mines Historical Society, 1980, pp. 18–23.
8 Noall, Cyril, *Levant: The Mine Beneath the Sea*, Truro, D. Bradford Barton, 1970, pp. 107–17.
9 Sharpe, Adam (ed.), *Engine House Assessment: Mineral Tramways Project*, Truro, Cornwall Archaeological Unit and Kerrier Groundwork Trust, 1991, pp. 22–40.
10 Palmer, Marilyn and Neaverson, Peter, 'The Comparative Archaeology of Tin and Lead Dressing in Britain during the Nineteenth Century', *Bulletin of Peak District Mines Historical Society*, vol. 10, no. 6, Winter 1989, pp. 316–53.
11 Bick, David, *The Old Metal Mines of Mid-Wales, Part 6: A Miscellany*, Newent, The Pound House, 1991, pp. 42–9.
12 Palmer, Marilyn and Neaverson, Peter, 'Nineteenth Century Tin and Lead Dressing: A Comparative Study of Field Evidence', *Industrial Archaeology Review*, vol. XII, no. 1, Autumn 1989, pp. 20–9.
13 Palmer, Marilyn and Neaverson, Peter, *The Basset Mines of Cornwall: Their History and Industrial Archaeology*, British Mining No. 32, Sheffield, Northern Mine Research Society, 1987.
14 Hughes, Stephen and Reynolds, Paul, *A Guide to the Industrial Archaeology of the Swansea Region*, Ironbridge, Association for Industrial Archaeology, 1989, pp. 11–14.
15 Raistrick, Arthur, *The Lead Industry of Wensleydale and Swaledale: Volume 2 The Smelting Mills*, Hartington, Moorland, 1975, pp. 35–48, 73–8, 105–8.
16 Day, Joan, *Bristol Brass: The History of the Industry*, Newton Abbot, David & Charles, 1973, p. 26ff.

17 Hayes, Raymond H., *A History of Rosedale*, Helmsley, North York Moors National Park, 1985, pp. 60–89.
18 Cleere, Henry and Crossley, David, *The Iron Industry of the Weald*, Leicester, Leicester University Press, 1985.
19 Riden, Philip, 'The Ironworks at Alderwasley and Morley Park', *Derbyshire Archaeological Journal*, vol. CVIII, 1988, pp. 77–107.

5 Clothing the people

1 Defoe, Daniel, *A Tour Through the Whole Island of Great Britain, 1724–1726*, vol. II, letter VIII, ed. G.D.H. Cole and D.C. Browning, London, J.M. Dent, 1962, p. 194.
2 Information and Local Interest Trails published by the Saddleworth Historical Society.
3 Giles, Colum and Goodall, Ian H., *Yorkshire Textile Mills 1770–1930*, London, HMSO, 1992, p. 221.
4 Jenkins, J. Geraint, *The Welsh Woollen Industry*, Cardiff, National Museum of Wales, 1969.
5 Park, Brian A., *The Woollen Mill Buildings in the Hillfoots Area*, Stirling, Forth Naturalist and Historian, 1984.
6 Durie, Alastair J., *The Scottish Linen Industry in the Eighteenth Century*, Edinburgh, John Donald, 1979, pp. 14–20.
7 McCutcheon, Alan, *Wheel and Spindle*, Belfast, Blackstaff Press, 1977, p. 51.
8 Timmins J.G., *Handloom Weavers' Cottages in Central Lancashire*, Lancaster, University of Lancaster, 1977.
9 Watson, Mark, *Jute and Flax Mills in Dundee*, Tayport, Hutton Press, 1990.
10 Chapman, Stanley D., 'The Arkwright Mills – Colquhouns's Census of 1788 and Archaeological Evidence', *Industrial Archaeology Review*, vol. VI, no. 1, Winter 1981–2, pp. 5–27.
11 Chapman, Stanley D., *The Early Factory Masters*, Newton Abbot, David & Charles, 1967; Fitton, R.S. and Wadsworth A.P., *The Strutts and the Arkwrights 1758–1830: A Study of the Early Factory System*, Manchester, Manchester University Press, 1958.
12 Butt, John (ed.), *Robert Owen, Prince of Cotton Spinners*, Newton Abbot, David & Charles, 1971.
13 Greatrex, Nan, 'The Robinson Enterprises at Papplewick, Nottinghamshire', *Industrial Archaeology Review*, vol. IX, no. 1, Autumn 1986, pp. 37–56, and vol. IX, no. 2, Spring 1987, pp. 119–39.
14 Dickinson, T.C., *Lancashire under Steam*, Preston, Lancashire County Council, 1984, p. 32.
15 Fitzgerald, Ron, 'The Development of the Cast Iron Frame in Textile Mills', *Industrial Archaeology Review*, vol. X, no. 2, Spring 1988, pp. 127–45; Jones, Edgar, *Industrial Architecture in Britain, 1750–1939*, London, Batsford, 1985; Holden, Roger N., 'Structural Engineering in the Lancashire Cotton Spinning Mills 1850–1914: The Example of Stott & Sons', *Industrial Archaeology Review*, vol. XV, no. 2, Spring 1993, pp. 160–76.
16 Gurr, Duncan and Hunt Julian, *The Cotton Mills of Oldham*, Oldham, Oldham Leisure Services, 1989 pp. 4–11; Williams, Mike with Farnie, D.A., *Cotton Mills in Greater Manchester*, Preston, Carnegie Publishing, 1992.
17 Defoe, Daniel, *A Tour Through England and Wales*, vol. II, letter VIII, ed. G.D.H. Cole and D.C. Browning, London, J.M. Dent, 1962, p. 156.

18 Calladine, Anthony and Fricker, Jean, *East Cheshire Textile Mills*, London, HMSO, 1993.
19 Prest, John, *The Industrial Revolution in Coventry*, Oxford, Oxford University Press, 1960, pp. 96–135.
20 Felkin, W., *A History of the Machine Wrought Hosiery and Lace Manufactures*, 1867, reprinted Newton Abbot, David & Charles, 1967.
21 Palmer, Marilyn and Neaverson, Peter, *Industrial Landscapes of the East Midlands*, Chichester, Phillimore, 1992, pp. 5, 16, 171–8, 183–6.
22 ibid.: pp. 15–16, 132–52.
23 Twilley, Royston and Wilks, Michael (eds), *The River Wandle*, Sutton, Borough of Sutton Libraries, 1974, pp. 20–1.
24 Freeman, Charles, *Luton and the Hat Industry*, Luton, Luton Museum, 1953.
25 Palmer and Neaverson, pp. 178–82.

6 Building and servicing the community

1 Ritchie-Noakes, Nancy, *Liverpool's Historic Waterfront*, London, HMSO, 1984, p. 58.
2 Stanier, Peter, 'Granite Quarrying in Devon and Cornwall, Part One: 1800–1910', *Industrial Archaeology Review*, vol. VII no. 2, Spring 1985, pp. 171–89.
3 Ewans, M.C., *The Haytor Granite Tramway and Stover Canal*, Newton Abbot, David & Charles, 1977, pp. 18–29.
4 Lindsay, Jean, *A History of the North Wales Slate Industry*, Newton Abbot, David & Charles, 1974, p. 88.
5 Lewis, M.J.T. and Denton, J.H., *Rhosydd Slate Quarry*, Shrewsbury, The Cottage Press, 1974, reprinted Mold, Adit Publications, 1994.
6 Tucker, D.G., 'Millstone making in the Peak District of Derbyshire', *Industrial Archaeology Review*, vol. VIII, no. 1, Autumn 1985, pp. 42–58.
7 Marshall, Gary, Palmer, Marilyn and Neaverson, Peter, 'The History and Archaeology of the Calke Abbey Limeyards', *Industrial Archaeology Review*, vol. XIV, no. 2, Spring 1992, pp. 145–76.
8 Trueman, Michael R.G., 'The Langcliffe Quarry and Limeworks', *Industrial Archaeology Review*, vol. XIV, no. 2, Spring 1992, pp. 126–44.
9 Smith, John R., *Cornwall's China Clay Heritage*, Truro, Twelveheads Press, 1992.
10 Baker, Diane, *Potworks*, London, Royal Commission on the Historical Monuments of England, 1991.
11 Torrington, Viscount, *The Torrington Diaries 1781–1794: A Tour to the North (1792)*, vol. III, London, Eyre & Spottiswoode, 1936, p. 127.
12 Spavold, J., 'The Sanitary Pottery Industry of South Derbyshire', *Industrial Archaeology Review*, vol. V, no. 2, Spring 1981, pp. 143–54.
13 Torrington, op. cit., p. 82.
14 Calladine, Anthony and Fricker Jean, *East Cheshire Textile Mills*, London, Royal Commission on the Historical Monuments of England, 1993, pp. 147–50.
15 Gaskell, S. Martin, *Building Control: National Legislation and the Introduction of Local Bye Laws in Victorian England*, London, British Association for Local History, 1983.
16 Trainor, Richard, 'Peers on an Industrial Frontier: The Earls of Dartmouth and of Dudley in the Black Country, c. 1810 to 1914', in David Cannadine

(ed.), *Patricians, Power and Politics in Nineteenth Century Towns*, Leicester, Leicester University Press, 1982, p. 75.
17 Mathias, Peter, *The First Industrial Nation*, London, Methuen, 1969, appendix: tables 1 and 3.
18 Phillips, J.F., *Town and Village in the Nineteenth Century, Nottingham and Nottinghamshire Villages*, Nottingham, University of Nottingham, 1972.
19 Binnie, G.M., *Early Dam Builders in Britain*, London, Thomas Telford, 1987.
20 Leapman, Michael (ed.), *The Book of London*, London, Guild Publishing, n.d. (*c.* 1990), pp. 178–9; Weightman, Gavin, *London River*, London, Guild Publishing, 1990, pp. 114–29.
21 Binnie, G.M., *Early Victorian Water Engineers*, London, Thomas Telford, 1981.
22 Darby, H.C., *The Changing Fenland*, Cambridge, Cambridge University Press, 1983; Hills, Richard L., *Machines, Mills and Uncountable Costly Necessities: A Short History of the Drainage of the Fens*, Norwich, Goose, 1967; Summers, Dorothy, *The Great Level: A History of Drainage and Land Reclamation in the Fens*, Newton Abbot, David & Charles, 1976.
23 Carr, R.J.M. (ed.), *Dockland: An Historical Survey of Life and Work in East London*, London, North East London Polytechnic, 1986, pp. 214–16, 234, 254.
24 Tucker, D.G., 'Refuse Destructors and Their Use for Generating Electricity: A Century of Development', *Industrial Archaeology Review*, vol. II, no. 1, Autumn 1977, pp. 5–27.
25 Irlam, Geoffrey A., 'Electricity Supply at Cragside', *Industrial Archaeology Review*, vol. XI, no. 2, Spring 1989, pp. 187–95.
26 Tucker, D.G., 'Hydro-electricity for Public Supply in Britain: 1881–1894', *Industrial Archaeology Review*, vol. I, no. 2, Spring 1977, pp. 126–63.

7 Moving around: roads, rivers, canals and railways

1 Hey, David, *Packmen, Carriers and Packhorse Roads*, Leicester, Leicester University Press, 1980.
2 Bird, Anthony, *Roads and Vehicles*, Harlow, Longman, 1969; Wright, G.N., *Turnpike Roads*, Princes Risborough, Shire, 1992.
3 Young, Arthur, *A Six Month Tour through the North of England*, vol. I, 1771, reprinted New York, Augustus Kelley, 1967, p. 349; ibid.: vol. IV, p. 428.
4 Rolt, L.T.C., *Thomas Telford*, London, Longman, 1959.
5 Crocker, Glenys (ed.), *A Guide to the Industrial Archaeology of Surrey*, Ironbridge, Association for Industrial Archaeology, 1990, pp. 7, 26, 39.
6 Rolt, L.T.C., *Navigable Waterways*, London, Longman, 1969.
7 Lewis, M.J.T., *Early Wooden Railways*, London, Routledge & Kegan Paul, 1970.
8 Grenter, Stephen, 'A Wooden Waggonway Complex at Bersham Ironworks, Wrexham', *Industrial Archaeology Review*, vol. XV, no. 2, Spring 1993, pp. 195–207.
9 Awdry, Rev. A.W. (ed.), *Industrial Archaeology of Gloucestershire*, Dursley, Gloucestershire Society for Industrial Archaeology, 3rd edn, 1983, pp. 10–13.
10 Hadfield, Charles, *The Canals of South Wales and the Border*, Newton Abbot, David & Charles, 1967.

11 Cook, R.A. and Clinker, C.R., *Early Railways Between Abergavenny and Hereford*, Oakham, Railway and Canal Historical Society, 1984; Rattenbury, G., *Tramroads of the Brecknock and Abergavenny Canal*, Oakham, Railway and Canal Historical Society, 1980.
12 Clinker, C.R. and Hadfield, Charles, 'The Ashby-de-la-Zouch Canal and its Railways', *Transactions of the Leicestershire Archaeological and Historical Society*, 1958, reprinted Bristol, Avon Anglia, 1978; Marshall, Gary, Palmer, Marilyn and Neaverson, Peter, 'The History and Archaeology of the Calke Abbey Limeyards', *Industrial Archaeology Review*, vol. XIV, no. 2, Spring 1992, pp. 145–76.
13 Rimmer, A., *The Cromford and High Peak Railway*, Lingfield, Oakwood Press, 1985.
14 Pugsley, Sir Alfred (ed.), *The Works of Isambard Kingdom Brunel*, London and Bristol, Institution of Civil Engineers and University of Bristol, 1976; Rolt, L.T.C., *Isambard Kingdom Brunel*, London, Longmans Green, 1957.
15 Smiles, Samuel, *Lives of the Engineers*, 1862, reprinted Newton Abbot, David & Charles, 1968, p. 481.
16 Biddle, Gordon and Nock, O.S., *The Railway Heritage of Britain*, London, Michael Joseph, 1983.
17 ibid.
18 Caffyn, Lucy, *Workers' Housing in West Yorkshire, 1750–1920*, London, HMSO, 1986, pp. 70–2.
19 Binney, Marcus and Pearce, David, (eds), *Railway Architecture*, London, Bloomsbury Books, 1979, pp. 118–39.
20 Andrew, J.H., 'The Smethwick Engine', *Industrial Archaeology Review*, vol. VIII, no. 1, 1985, pp. 7–27; Andrew, J.H., 'Canal Pumping Engines', *Industrial Archaeology Review*, vol. XV, no. 2, 1993, pp. 140–59.
21 Hadfield, Charles, *The Canals of the West Midlands*, Newton Abbot, David & Charles, 1985.
22 Conway-Jones, Hugh, *Gloucester Docks: An Illustrated History*, Gloucester, Alan Sutton, 1984.
23 Coad, Jonathan, *Historic Architecture of the Royal Navy*, London, Gollancz, 1983; MacDougall, Philip, *Royal Dockyards*, Princes Risborough, Shire, 1989.
24 Lord, John and Southam, Jem, *The Floating Harbour, a Landscape History of Bristol City Docks*, Bristol, Redcliffe Press, 1983.
25 Ritchie-Noakes, Nancy, *Liverpool's Historic Waterfront*, London, HMSO, 1984; Jarvis, Adrian, *Docks of the Mersey*, London, Ian Allan, 1988.
26 Hume, John R., *The Industrial Archaeology of Glasgow*, Glasgow, Blackie, 1974.
27 Carr, R.J.M., *Dockland: An Illustrated Historical Survey of Life and Work in East London*, London, North East London Polytechnic, 1986.

8 The Industrial landscape: past, present and future

1 Harris, J.R., 'Industrial Espionage in the Eighteenth Century' *Industrial Archaeology Review*, vol. VII, no. 2, 1985, pp 127–38.
2 Moir, Esther, *The Discovery of Britain*, London, Routledge & Kegan Paul, 1964.
3 Torrington, Viscount, *The Torrington Diaries 1781–1794: A Tour to the North (1792)*, vol. III, London, Eyre & Spottiswoode, 1936, pp. 81–2.
4 ibid, vol. II, pp. 195–6.

5 Britton, J., *A Topographical and Historical Description of the County of Norfolk*, vol. II, London, 1813, p. 344.
6 Svedenstierna, Eric T., *Svedenstierna's Tour of Great Britain 1802–3*, ed. M.W. Flinn, Newton Abbot, David & Charles, 1973, pp. 54–5.
7 Webb, D.C., *Observations and Remarks during Four Excursions made to Various Parts of Great Britain in the Years 1810–1811*, London, 1812, pp. 186–90.
8 Bingley, Rev. W., *A Tour Round North Wales . . . during the Summers of 1798 and 1801*, vol. I, London, 1804, pp. 309–10.
9 Black's *Picturesque Tourist and Road and Railway Guide through England and Wales*, Edinburgh, A. & C. Black, 1850, p. 250.
10 Burritt, Elihu, *Walks in the Black Country and its Green Border-Land*, London, Sampson Low, 1868, pp. 179–80.
11 Ross, Michael, *Planning the Heritage*, London, Chapman and Hall, 1991.
12 Clark, Catherine, *English Heritage Book of Ironbridge Gorge*, London, Batsford, 1993.
13 Hewison, Robert, *The Heritage Industry*, London, Methuen, 1987.
14 Binney, Marcus and Watson-Smyth, Marianne, *The SAVE Britain's Heritage Action Guide*, London, Collins & Brown, 1991; Binney, Marcus, Machin, Francis and Powell, Ken, *Bright Future*, London, SAVE Britain's Heritage, 1990.
15 Giles, Colum and Goodall, Ian H., *Yorkshire Textile Mills 1770–1930*, London, HMSO, 1992; Williams, Mike with Farnie, D.A., *Cotton Mills of Greater Manchester*, Preston, Carnegie Publishing, 1992; Calladine, Anthony and Fricker, Jean, *East Cheshire Textile Mills*, London, Royal Commission on the Historical Monuments of England, 1993.

Index

Note: Page numbers in bold denote illustrations.

For Product Safety Concerns and Information please contact our EU representative GPSR@taylorandfrancis.com
Taylor & Francis Verlag GmbH, Kaufingerstraße 24, 80331 München, Germany

www.ingramcontent.com/pod-product-compliance
Ingram Content Group UK Ltd.
Pitfield, Milton Keynes, MK11 3LW, UK
UKHW021302190926
475701UK00009B/616
* 9 7 8 0 4 1 5 5 1 3 4 5 6 *